For George and Kim

Table of Contents
Dr. Phil's R75 HANDBOOK
©2016

CHAPTER 1: IC-R75 COOKBOOK AND MODIFICATIONS

1.1 The R75 Cookbook (Using the ICOM IC-R75.)
1.2 Phil's Soldering 101
1.3 Phil's Fidelity Increase Mod
1.4 Phil's Fidelity Increase Mod (Application Notes)
1.5 Phil's Cheap AM Filter Mod
1.6 PhilMod ECSS Volume Increase
1.7 Pete's MW Attenuation Removal Mod
1.8 Display Frequency Calibration
1.9 Voice Synthesizer Volume Adjustment
1.10 DrPhilSCAN Cookbook (Easy Tuning Mode in 2002)

CHAPTER 2: SHORTWAVE RECIEVER ARTICLES

2.1 E1 versus R75 (Indoor Shootout From 2007)
2.2 R8B versus R75
2.3 SSB Cookbook (Single Sideband for Portables)
2.2 DE1103 Cheat Sheet (Using your DE1103.)
2.3 The DE1102 Mini-Cookbook
2.4 KA2100: Superior by Design
2.5 FRG-100 SWL Tips (Using the Yaesu FRG-100.)
2.6 Eton E1 SWL Tips (Using the Eton E1.)
2.7 Eton E1 Flaw List (Design flaws of the Eton E1.)
2.8 E1XM versus Sat800 (Comparing radios with SAM.)
2.9 E1 versus R75
2.10 R8B versus R75

CHAPTER 3: SHORTWAVE RECIEVER BUYING GUIDES

3.1 Portable Shortwave Radio Guide
3.2 Phil's SW Radio Picks
3.3 Phil's SW Radio Buying Guide
3.4 Phil's Tabletop Guide 2005
3.5 Phil's Portable Guide 2005
3.6 Tabletop Receiver Notes
3.7 Mid-70's Panasonic Portables

This page intentionally blank.

1.1 The R75 Cookbook
Using the ICOM IC-R75.
©2002

1. INTRODUCTION

This cookbook is a crash course in using the R75. It contains tricks to improve reception and other useful information. Using the buttons and knobs cannot hurt the receiver; all cautions below are simply to prevent getting blasted by excessive volume. Sections include: Fundamentals, Tricks, Scanning, Memories, and Computer Related. I would like to thank those who shared their tricks with me. The R75 community is our greatest asset.

2. FUNDAMENTALS

2.1 WHAT IS ECSS?

ECSS is an acronym for Exalted Carrier Selectable Sideband. ECSS is the tuning of AM stations as if they were SSB. ECSS reduces selective-fading related distortion, and allows choosing the sideband (LSB or USB) with the least interference. The R75 has the stability (1 ppm) and fine tuning steps (1 Hz) necessary for excellent ECSS reception. To display 1 Hz resolution press 'TS' until the inverted triangle disappears and then depress 'TS' 2 seconds.

2.2 SETUP MODE ESSENTIALS

- Press 'SET' then press 'UP' or 'DN' until 'RF/SQL' is displayed.
- Turn the tuning dial to the right until 'r5' (aka RS) is displayed.
- Press 'SET' again to exit.

This allows manual gain control and is **VERY IMPORTANT** to proper use of the tricks below. The left half of the RF/SQL knob is the RF gain control and the right half is the squelch control.

- Press 'SET' then press 'UP' or 'DN' until 'REC/RM' is displayed.
- Turn the tuning dial to the left until 'OFF' is displayed.
- Press 'SET' again to exit.

This will turn off the recorder remote and **STOP RELAY CLICKING**, an annoyance. You may also wish to change 'BK LIGHT' (backlight level) and 'BP LVL' (beep level).

2.3 FILTER SETUP

I use the following (9 MHz & 455 kHz) filter settings on my stock R75:

- AM normal = 15 kHz & 6 kHz
- AM narrow = 2.4 kHz & 2.4 kHz
- SSB normal = 2.4 kHz & 2.4 kHz
- SSB wide = 15 kHz & 6 kHz (for Robert's ECSS Fine Tuning Trick below)
- FM normal = 15 kHz & 15 kHz

Enter the filter setup mode by depressing 'FIL' 2 seconds, then use 'UP' and 'DN', and select filter values using the tuning knob. The 'N' stands for 'narrow' and the 'W' stands for 'wide'. SSB wide CANNOT be set as per above unless 'EXP' is first set to 'oN'. Press 'FIL' again to exit. The stock filters are quite good and tricks may make additional filters unnecessary.

2.4 SETTING THE CLOCK

Set the R75's clock to UTC or Coordinated Universal Time. UTC is available from the US Naval Observatory's Master Nuclear Clock website: http://tycho.usno.navy.mil/cgi-bin/timer.pl

- Press 'CLOCK'.
- Depress 'SET' 2 seconds (time will flash).
- Enter a 4-digit military time (or) Use the tuning dial.
- Press 'ENT' to accept (or) Press 'CLR' to cancel.
- Press 'CLOCK' again to exit.

2.5 NOISE CONTROL

For pulsating (ignition, lamp, neon) noise use the noise blanker (NB). For heterodynes (whistles) especially while tuning on SSB use the automatic notch filter (ANF). For atmospheric (static) noise use the noise reduction (NR), setting as follows:

- Depress 'NR' 2 seconds.
- Use the tuning knob to adjust.
- Press 'NR' when done.

A NR value between 3 and 5 usually works best. Engaging the NB without pulsating noise being present will cause interference.

2.6 GENERAL RECEPTION TIPS

Frequencies below 13 MHz work better at night while those above 13 MHz work better in the day. Ham radio operators use LSB below 10 MHz and USB above 10 MHz. Maritime, military, and aeronautical uses USB. Due to gray-line reception listening will be enhanced from a half hour before to a half hour after your local sunrise or sunset.

2.7 INEXPENSIVE INDOOR ANTENNAS AND LOOPS

I use a 50-foot roll of white wire wrap attached to the RED random-wire push connector terminal in back. At 30-guage size it is nearly invisible against white walls (or the ceiling) and the cost at Radio Shack is only $4.99 (#278-502). Loop antennas are excellent for indoor reception. Radio waves contain both electrical and magnetic components. The former contains most of the local noise. A loop antenna responds mostly to the latter. Loops are also directional (figure-8) in nature.

2.8 "SOMETHING IS WRONG" LIST FOR BEGINNERS

No sound? Check this list first:

- Set the RF gain to 12 o'clock.
- Turn up the volume.
- Turn the attenuator OFF.
- Set both PBTs to 12 o'clock.
- Turn the AGC ON (press momentarily).
- Depress 'ANT' 2 seconds (select the correct antenna).

If the tuning dial is not working press the black button under the 'DN' button to release the 'LOCK' mode. The lock mode is intended to prevent accidental changes. Increase tuning dial tension by moving the lever below the dial. This lever should be pulled slightly outward (toward you) and then slid to the left. The R75 has a mute circuit so quiet could denote audio overload. The R75 is so sensitive that if the wrong antenna (no antenna) is selected you may still hear stations faintly. If you can only hear the strongest stations try depressing 'ANT' for 2 seconds!

3. TRICKS

CAUTION: Start with the volume low and try these tips without headphones first.

Indispensable Tricks

3.1.1 AUDIO ENHANCEMENT TRICKS

- Purchase computer speakers for program listening ($30).
- Purchase earbuds for DXing ($5).
- Possibly purchase the Koss EQ50 equalizer ($20).

This first trick is 'cheating' (hardware) but the stock R75 speaker is small and upgrading here will help sound quality tremendously. Attach amplified speakers to 'REC' and un-amplified speakers to 'EXT SP'.

3.1.2 BRYANT'S AGC SHUTOFF TRICK

- Decrease RF gain. **IMPERATIVE**
- Turn the AGC OFF (depress 'AGC' 2 seconds).
- Increase RF gain until s-meter drops to 9-20 dB.
- Adjust volume.

This trick prevents AGC pumping by allowing the minimum gain for a signal. It also stops AGC and RF back-mixing. Using as little gain as possible is generally a good policy: turn off the preamps or engage the attenuator whenever possible. This is very important!

CAUTION: Never tune or scan with the AGC OFF! Press 'AGC' again to turn it back on.

3.1.3 Bruce's Fade Reducing Trick:

- Decrease RF gain until s-meter starts rising.

This trick will decrease fading. Use it with the AGC on.

AM Tricks

3.2.1 PHIL'S AM DETUNING TRICK

- Using the AM mode engage dual 2.4 kHz filters.
- Detune the station exactly +/- 1.2 kHz (aka half the filter value).

This trick increases fidelity over just using the narrow filter on AM and may cut out adjacent sideband interference. Example: For 9010 kHz AM engage the narrow filter and tune either 9011.2 kHz or 9008.8 kHz. Using the 6 kHz AM filter, detune down by 0 to 2.5 kHz for LSB-only or up by 0 to 3.5 kHz for USB-only. Make sure the 'S' in 'S-AM' is not blinking (disengaged).

3.2.2 ROBERT'S ECSS FINE TUNING TRICK

- Set mode to SSB for an AM signal (ECSS).
- Tune until the signal sounds natural.
- Engage the 6 kHz filter.
- Tune until fluttering modulation stops.
- Engage the 2.4 kHz filter.

This **GREAT** trick helps make ECSS fine-tuning easy and fun! To set the wide SSB filter to 6 kHz see 'Filter Setup' above.

3.2.3 PINPOINTING AM STATION TRICKS

- Engage dual 2.4 kHz filters on AM.
- (or) Switch to SSB and set tuning steps to 5 kHz.
- (or) Confirm the signal exists on both LSB and USB.

These tricks allow finding the station's carrier frequency since the R75's AM chip is capable of a wide locking range with the 6 kHz filter engaged. After finding the center switch back to the 6 kHz filter for better fidelity. To set tuning step depress 'TS' 2 seconds. I use 5 kHz on AM and 1 kHz on SSB.

3.2.4 SAM USAGE TRICK

- Press 'SET' then press 'UP' or 'DN' until 'SAM SW' is displayed.
- Using the tuning knob select 'En' (aka selectively use SAM).
- Press 'SET' again to exit.
- Set mode to SAM (press 'AM' if necessary).
- Increase the RF gain until the 'S' in 'S-AM' is not blinking (activated).

This trick will allow the S-AM mode to work "as well as can be expected".

SSB Tricks

3.3.1 SSB FILTER BANDWIDTH TRICK

- Turn the two PBTs in opposite directions.

This trick will allow SSB bandwidths of ~300 Hz to 2400 Hz. For ham reception move the PBTs in opposite directions one white mark. The inner knob is the 455 kHz IF. Moving the PBTs together (same direction) causes bandpass shifting.

3.3.2 SSB TONE CONTROL TRICKS

- Detune the station up to +/- 30 Hz.
- (or) Turn one of the PBT controls to the left or right.

These tricks allow good tone control on SSB. On USB introduce bass by either increasing the frequency or turning the inner PBT knob to the right.

3.3.3 WHICH DIRECTION TO TUNE ON THE OTHER SIDEBAND TRICK

- After switching to LSB, tune higher in frequency.
- After switching to USB, tune lower in frequency.

This trick helps pinpoint a signal after changing sidebands. Example: If a heterodyne is heard on LSB 14243 kHz, change to USB and tune **lower** to find a signal at USB 14240 kHz.

3.3.4 PHIL'S "I USE ECSS EXCLUSIVELY, CAN AM BE DISABLED?" TRICK

Note: This trick requires at least one missing optional filter.

- Depress 'FIL' 2 seconds.
-
- Press 'UP' twice. ('9M' 'N' is displayed)
- Turn the dial until 'OFF' is displayed.

- Press 'UP' once. ('9M' 'W' is displayed)
- Turn the dial until 'OFF' is displayed.

- Press 'UP' once. ('EXP oFF' is displayed)
- Turn the dial until 'ON' is displayed.

(If a 9 MHz optional filter **is present** use the **BOLD** instead.)
- Press 'UP' until 'oP1' (**'oP2'**) is displayed.
- Turn the dial until '100' (**'52A'**) is displayed.
- Press 'UP' until '9M' (**'455K'**) is displayed.
- Turn the dial until '0.5' is displayed.

- You should hear NO SOUND!
- Press 'FIL' again to exit.

This trick prevents high volume in the event of accidentally switching from ECSS to AM. The CPU is told to use an optional filter that is missing; therefore, when AM is engaged an open circuit exists and no signal gets through. A less invasive albeit less effective method is to set AM to dual 2.4 kHz filters. This trick is also capable of muting FM.

Miscellaneous Tricks

3.4.1 ROBERT'S ANTENNA MUTE TRICK

- Depress 'ANT' 2 seconds (switch to a missing antenna).

This trick will mute the audio when only one antenna is present. This avoids having to adjust the volume control back to previous levels following an interruption.

3.4.2 TRANSMITTER VERSUS RECEIVER DISTORTION DIAGNOSTIC TRICK

- Turn PreAmps and attenuator OFF.
- Tune till distortion is 9 dB on s-meter.
- Press 'ATT' to engage the attenuator.
- Note the s-meter reading.

This trick is diagnostic: if the s-meter reads ~6 dB suspect the distortion is at the transmitter; however, if it is **much less than** 6 dB then suspect the receiver.

3.4.3 BRUCE'S SUPER DXING TRICK

- Set mode to SSB.
- Decrease the volume. **IMPERATIVE**
- Turn AGC OFF (depress AGC 2 seconds).
- Turn the RF gain to maximum (12 o'clock).
- Select PreAmp '2'.
- Spin dial looking for DX (distance reception).

This trick allows high gain for DXing but use with **CAUTION** (can be very loud). **WARNING**: Use with extreme care (use stock speaker, NOT headphones).

3.4.4 RF VOLUME TRICK

- Turn RF gain to zero. **IMPERATIVE**
- Increase volume fairly high.
- Use the RF gain as volume control

EFFECT: May help reception but use with **CAUTION** (can be very loud). **WARNING**: Use with extreme care (use stock speaker, NOT headphones).

4. SCANNING

4.1 SETUP MODE SCANNING OPTIONS: OPTIONAL

- Press 'SET' then press 'UP' or 'DN' until 'SCN RS' [Scan Resume] is displayed.
- Turn the dial to the right for 'ON' or to the left for 'OFF'.
- Press 'SET' again to exit.

'ON' means scanning resumes after stopping on a signal or when the signal disappears. 'OFF' means scanning stops and must be restarted manually.

- Press 'SET' then press 'UP' or 'DN' until 'SCN SPd' [Scan Speed] is displayed.
- Turn the dial to the right for 'HI' or to the left for 'LO'.
- Press 'SET' again to exit.

This changes the scan speed to high (HI) or low (LO).

4.2 REVIEW OF THE FIVE SCANNING MODES

Note: Pressing 'V/M' toggles between the VFO and MEMORY modes.
Note: Press 'SCAN' again to stop any of the scans below.

4.2.1 Programmed Scan - moves from P1 to P2 starting at the VFO frequency if possible.

- Turn the SQL (squelch) to the right of 12 o'clock.
- In VFO mode, press 'SCAN'.

Note: P1 and P2 are special memories located above '99' and below '1'.
Note: Setting P1 to 30 kHz and P2 to 60 MHz ensures starting at the VFO frequency!

4.2.2 Auto Memory Write Scan - as above but writes to memories 80 to 99.

- Turn the SQL (squelch) to the right of 12 o'clock.
- In VFO mode, press 'SCAN' then press 'M/W'.

4.2.3 Priority Watch - checks for activity on a memory channel every 5 seconds.

- Select memory to watch using 'UP' or 'DN'.
- Turn the SQL (squelch) to the right of 12 o'clock.
- In VFO mode, depress 'SCAN' for two seconds.

Note: A memory must be programmed to use priority watch.

4.2.4 Memory Scan - moves from memory 1 to 99.

- Turn the SQL (squelch) to the right of 12 o'clock.
- In MEMORY mode, press 'SCAN'.

4.2.5 Select Memory Scan - as above but only checks 'selected' channels.

- Turn the SQL (squelch) to the right of 12 o'clock.
- In MEMORY mode, press 'SCAN' then press 'SEL'.

Note: Usage was left out of the ICOM manual.
Note: Pressing 'SEL' while in the MEMORY mode toggles select on/off for each individual memory (notice that a small 's' appears next to each number if selected).

4.3 SCANNING TIPS

While using Programmed Scan or Auto Memory Write Scan make sure the AGC in ON!

While using Priority Watch, Memory Scan, or Select Memory Scan make sure that the PreAmp, Attenuator, AGC, and Antenna settings are the same or there will be reduced scanning speed, relay chatter, or Priority Watch will NOT be transparent. If this is not possible group similar settings sequentially.

5. MEMORIES

5.1 KEYPAD FREQUENCY ENTRY

- Type numbers followed by 'ENT', placing the decimal after the MHz place.
- It is not necessary to enter the MHz part on subsequent entries.
- Press 'CLR' to clear all digits.

Example: '14 ENT' turns into 14.000.000 MHz.
Example: '.7 ENT' after the above turns into 14.700.000 MHz.

5.2 MEMORY MANIPULATION

Selection of a Memory: Direct

- Enter via keypad the number of the memory (ex. 45).
- Press 'V/M'.

Selection of a Memory: Indirect

- Press 'V/M' to enter the MEMORY mode.
- Use 'UP' or 'DN' to go through memories.

Saving to Memory:

- Press 'UP' or 'DN' to select a memory.
- Depress 'M/W' (memory write) 2 seconds.

Clearing a Memory:

- Press 'V/M' to enter the MEMORY mode.
- Press 'UP' or 'DN' to select a memory.
- Depress 'CLR' 2 seconds.

Transfer a Memory to VFO:

- Press 'UP' or 'DN' to select a memory.
- Depress 'V/M' 2 seconds.

Note: Press 'V/M' to toggle between VFO and MEMORY modes.

ERASE ALL MEMORIES: WARNING

- Depress 'UP' and 'DN' while turning the power ON to ERASE ALL MEMORIES.

WARNING: THIS WILL ERASE ALL MEMORIES! Memories use non-volatile EEPROM and therefore survive power loss. The internal battery is for timekeeping.

6. ALPHANUMERIC LABELS

- Press 'V/M' to enter the MEMORY mode.
- Depress 'SEL' 2 seconds to enter the 'ALPHANUMERIC' mode.
- Press 'UP' or 'DN' to select a memory.
- Press 'ENT' to start (cursor blinking starts).
- Press 0 or 2-9 as needed (ex. to get a 'P' press '7' twice).
- Press period to enter a space.
- Use the tuning dial to move left or right through the digits.
- Press 'ENT' when done.

Note: Depress 'SEL' 2 seconds to exit the 'ALPHANUMERIC' mode.
Note: Names can be eight characters long but the last two must be numeric.
Note: While in the ALPHANUMERIC mode pressing 'TS' will show the frequency!
Note: It is easier to enter alphanumeric labels via a computer and transfer.

7. COMPUTER RELATED

7.1 DIGITAL TRANSMISSIONS

A computer can be used to decode digital transmissions without the need for additional filters. Connect the output labeled 'REC' to the computer sound card's input. To start try the 20-meter amateur band at night or on weekends, specifically: 14.025 to 14.150 MHz for BPSK/CW/RTTY, or 14.230 & 14.233 MHz for SSTV. Use the following software:

- **DigiPan** for BPSK or typed chat via radio. Includes an audio 'waterfall'.
http://www.digipan.net/

- **CWGet** for CW or Morse code.
http://www.dxsoft.com/micwget.htm

- **MMTTY** for RTTY or radio teletype (text) via radio.
http://www.qsl.net/mmhamsoft/mmtty/index.html

- **JVComm32** for SSTV or slow scan television (still pictures) via radio.
http://www.jvcomm.de/indexe.html

7.2 SOUND RELATED

Two very useful sound programs: (dead link? Try the Internet Archive Wayback Machine.)

- **SR5** is a digital filter and spectrum analyzer.
http://www.ar5.ndirect.co.uk/html/sr5.html

- **Scanner Recorder** is a digital audio recorder with squelch control (VOX).
http://www.davee.com/scanrec/index.html

7.3 COMPUTER CONTROL

ICOM's **RS-R75 Control Software** is $60; however, I recommend purchasing an RS232C (male/female DB9) cable from Radio Shack (#26-117B) and trying Mark Fine's shareware program first. His program is called **Smart ICOM Control 32** and usage after 60 days requires only a $60 registration fee. http://www.fineware-swl.com/sic.html

1.2 Phil's Soldering 101
©2002

1. SAFETY

Always wear ANSI Z87.1 compliant safety goggles while soldering. These goggles are made of polycarbonate, a type of plastic that resists shattering into small sharp pieces. Also be sure to unplug the soldering iron when done.

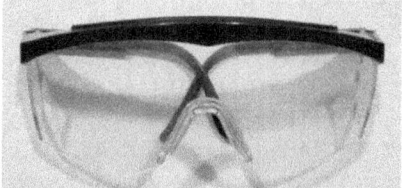

2. TOOLS

Obtain solder (A), soldering iron (B), wire-wrap (C), wire-wrap stripping tool (D), and scissors (E). A soldering iron holder and damp sponge are also needed (not shown). Solder should be of the 2 to 4 percent silver type. I highly recommend the ANTEX 12 Watt soldering iron model M/3U pictured below. It is inexpensive, small, and heats up in 45 seconds. It can be obtained from M. M. Newman Corporation, phone: (800) 777-6309. Make sure to also order a needle tip made of iron, specifically number 8-I. See the following website for more information: http://www.mmnewman.com/antexsi/antexpm.htm

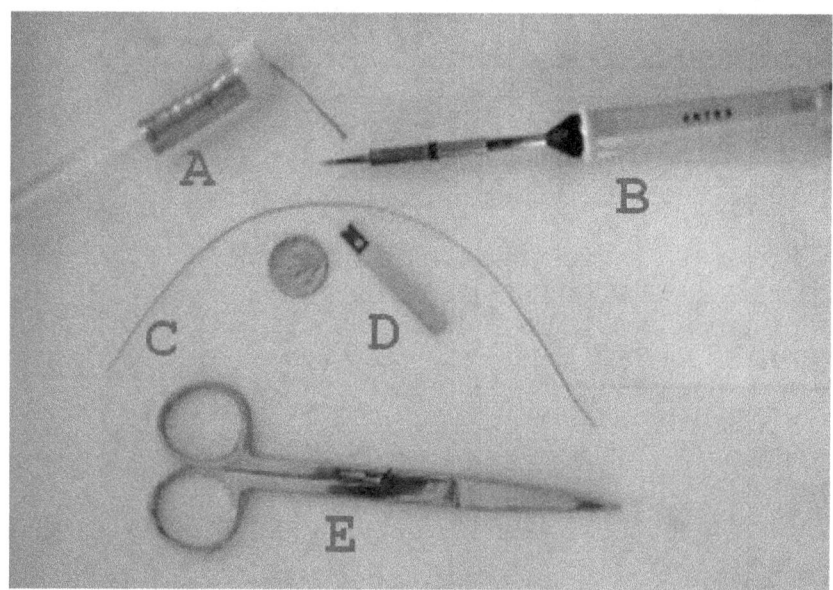

3. IRON PREPARATION

Add solder to the soldering iron. It may smoke (let it) as impurities burn off. Wipe the iron on the sponge. Do this until the iron tip looks shiny (red arrow) like a dime. The iron is now tinned.

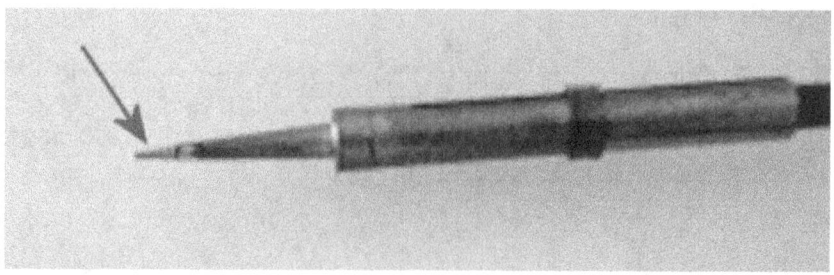

4. WIRE PREPARATION, PART A

Insert one end of the wire-wrap into the stripper tool (left arrow), pull towards the short end to strip (center) and then use the scissors to cut all but 1 mm of the bare wire (right arrow).

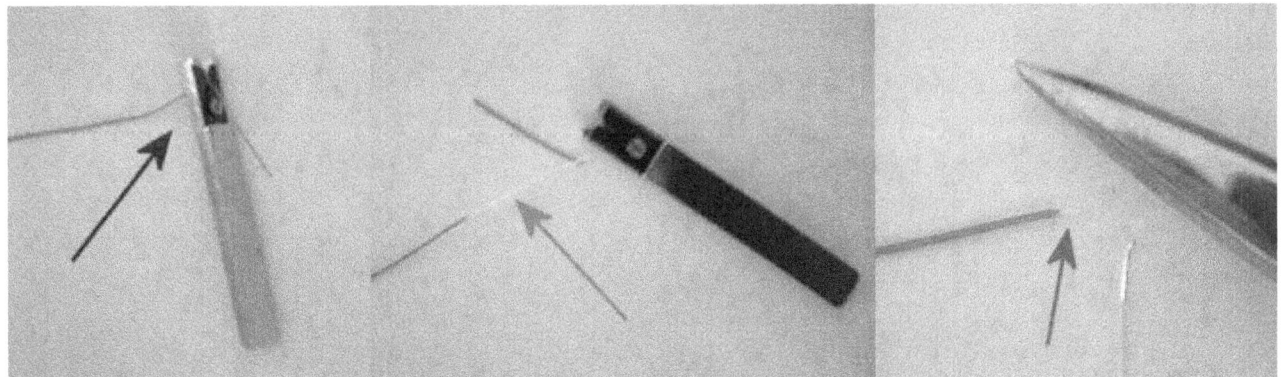

5. WIRE PREPARATION, PART B

Add solder to the iron. Dip the 1 mm of exposed wire into the solder (arrow) until very little solder is left on the wire (arrow). This may take several tries but is by far the most important step. Wipe all excess solder off the iron with the sponge and then heat the solder on the wire until it is no longer is jagged or horned (arrow). Notice how little solder is placed on the wire.

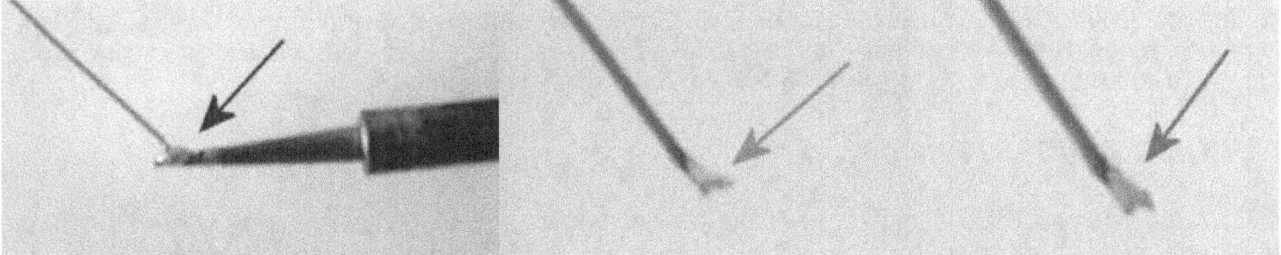

6. SOLDERING

Wipe all excess solder from the iron with the sponge. DO NOT add any additional solder (this is key to making a good clean surface-mount connection). Touch the wire (A) to the component (here a surface-mount capacitor indicated by the arrows). Using the side of the needle tip (arrow) apply heat for ~4 seconds while sandwiching the wire between the iron and component. If solder does not flow wait 30 seconds to allow for cooling and then re-apply heat. The connection should become shiny when the solder flows properly.

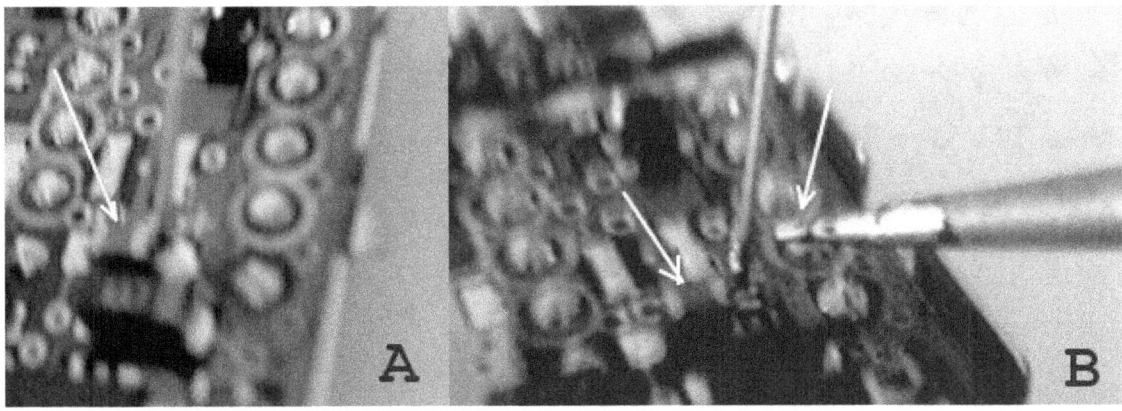

7. EXAMINATION

The soldering is now complete (A). Examine the connection visually. Do not be afraid to handle the connection, it should be very strong (B) and will not come apart unless you pull with significant force (more than you will accidentally apply). If the connection fails while handling, consider it a blessing because it was not a good connection. No need to lament, just re-apply heat and try again.

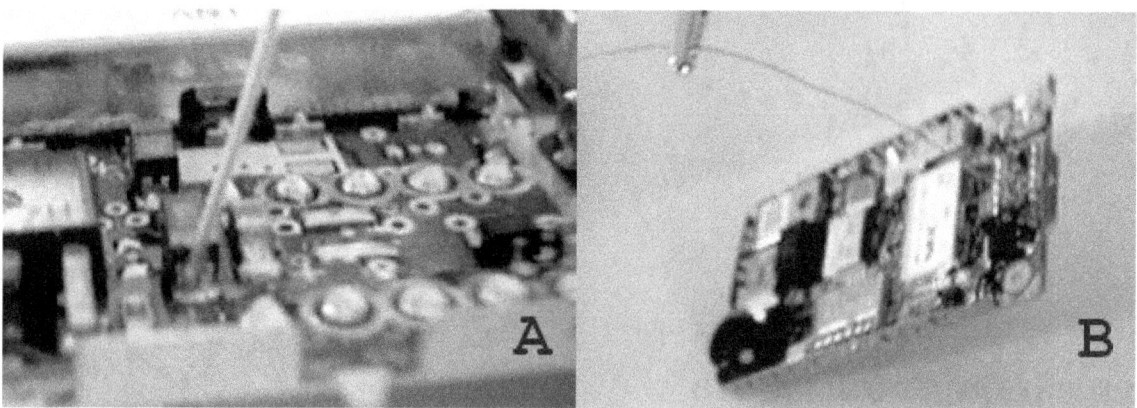

8. TIPS

Practice making ~15 connections like the one above on a junk surface-mount board like the old phone PCB above. Make sure to never overheat the board or any of its components. Heat kills electronics! Use a low wattage soldering iron (not a Radio Shack iron). Four seconds is not a hard and fast rule. What is important is solder flow. Once the solder is hot enough you will see it flow according to gravity and capillary action. A few moments after this, remove the soldering iron. Make sure to hold the wire in place for a few seconds while the solder cools.

9. FIXING A MESS, PART A

Lets say you have a big mess like the one below (arrow). Please note that kind of thing cannot happen when using the method described above. Desoldering braid (right arrow) to the rescue! Notice how the braid has been purposely flared at the end.

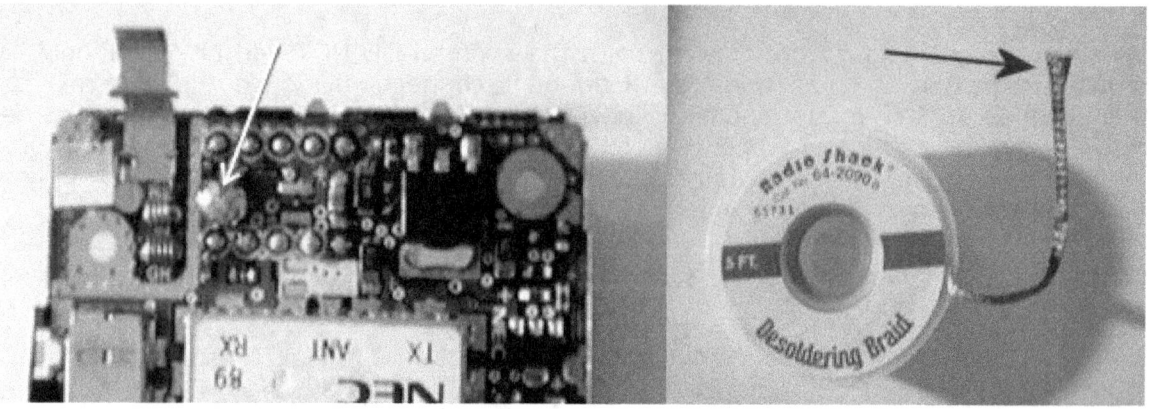

10. FIXING A MESS, PART B

Touch the braid to the mess (left arrow). Now press on the braid using the soldering iron (right arrow). It takes solder to flow solder (an engineer told me that back when I worked for a defense contractor); therefore, you may have to heat the solder directly for a moment before using this technique. DO NOT overheat the components. There is no rush.

11. FIXING A MESS, PART C

Notice how the solder gets sucked up into the braid (left arrow). Pull away when hot and most of the solder will be stuck to the braid (middle arrow). The mess is gone (right arrow). Small remaining amounts are often best removed by direct application of the iron (via capillary action).

12. FINAL EXAMINATION

If you have read up to here you can now solder to surface mount.

1.3 Phil's Fidelity Increase Mod
©2002

WARNING: Performing this mod will void your warranty and could **destroy** your radio.
WARNING: DO NOT perform this mod without some type of **eye protection**.
CAUTION: This mod takes soldering skill so please practice beforehand.
DISCLAIMER: The author is not responsible for any damage resulting from this mod.

1. ABSTRACT

Increase audio fidelity by soldering a separate 47k-ohm resistor in parallel with resistors R1212, R1213, R1214, and R1215. This is a modification for audiophiles (i.e. program and music listeners) who wish to enhance the already excellent AM (SAM) sound gained from Dr. Rado's Sync-AM and AM-AGC mods.

2. INTRODUCTION

The R75 block diagram shows that a 3 kHz low-pass filter exists between the audio-frequency select chip (IC1201) and the audio pre-amp. The audio bandwidth is purposely decreased to aid functioning of the optional UT-106 DSP. Unfortunately this attenuates the R75's high frequency audio response. Inspection of the schematics revealed that ICOM utilizes dual active two-pole low-pass filters. These two filters (shown below) employ op-amps (triangles below), specifically ¼ of the NJM2058 Quad Operational Amplifier chip each.

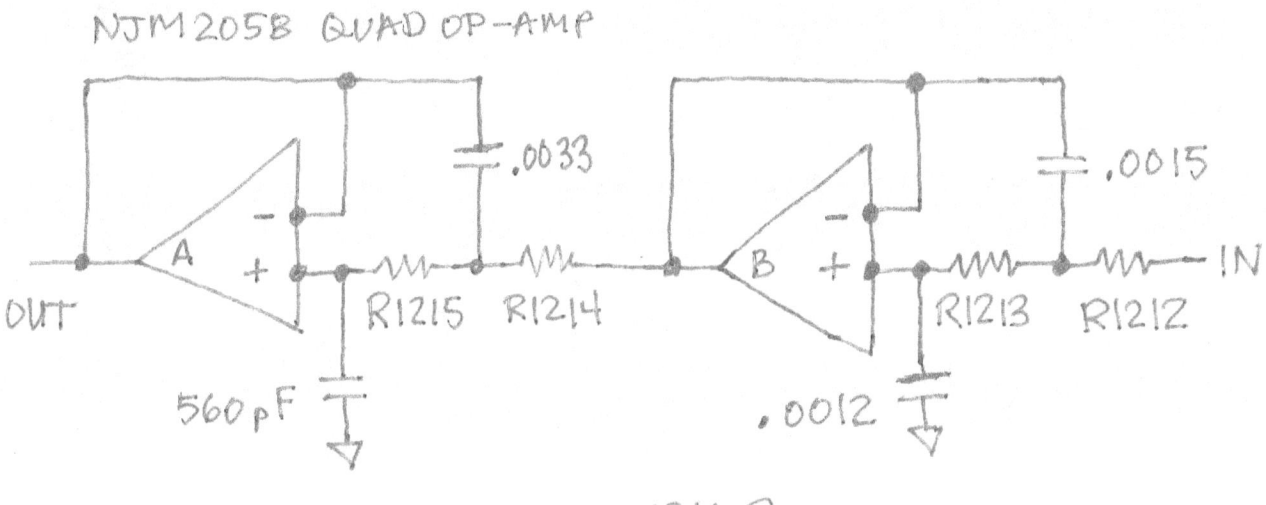

3. THEORY

The following equation was used for calculations: $4 * pi^2 * f^2 * R1 * R2 * C1 * C2 = 1$. Where f is the cutoff frequency, R1 & R2 are resistance values, and C1 & C2 are capacitance values. I calculated the stock 1211A (left op-amp above) filter as being set to 3001 Hz and the stock 1211B (right op-amp above) filter as being set to 3041 Hz.

The R75's audio fidelity can be increased by soldering a resistor in parallel with each of the four 39k-ohm resistors (shown in the schematic above) in the low-pass filters. Using some theoretical guidelines Ken experimentally determined the optimal value of the four parallel resistors to be 47k-ohms. Usage of these four 47k-ohm resistors in parallel with the stock 39k-ohm resistors yields 21.3k-ohms total effective resistance and will raise the bandwidth of the low-pass filters to 5493 Hz and 5566 Hz respectively. [Please note that as an alternative four 100k-ohm resistors in parallel (28k-ohms total effective resistance) will yield ~4200 Hz of bandwidth.]

It was originally theorized that the DSP used a 6000 Hz sample rate and that filtration was needed to eliminate signals above the Nyquist frequency to avoid aliasing distortions. However, these filters may have been added to avoid the noticeable loss of highs incurred by DSP usage. Engaging the DSP (with the NR or ANF button) returns the R75 to its unmodified sound.

Spectral analysis revealed an increase in frequency response from 3200 Hz in the stock unit to 4800 Hz in the modified unit. The result is unimpeded high fidelity sound from the R75. Music sounds fuller while using the 6 kHz filter and even more so using the pseudo dual-15 kHz filter setting. Usage of amplified external speakers with tone control is recommended.

4. WARNINGS AND PREPARATION

Please wear eye protection. Before you begin disconnect all cables ESPECIALLY the power supply and avoid static discharge by grounding yourself. Be careful that the soldering iron does not burn any cables! Be gentle when removing and putting back the power supply cable.

5. MATERIALS

To perform this modification you will need the following items:

- Four 47k-ohm resistors (Radio Shack sells a 5-pack for 99 cents).
- Thin insulated wire (ex. wire wrap) and electrical tape (to insulate exposed wire).
- A low-wattage needle-tipped soldering iron, holder, sponge, and solder.
- Safety glasses or goggles.

6. METHODS

- Take off the top cover by removing eight screws (see user manual for details).
- Locate the general region where NJM2058 op-amp is located from the picture below (see arrow and square).

Locate the four 39k-ohm resistors shown inside the circles below. Using a magnifier you will note the resistors have the marking '393' that stands for 39k-ohms. You can check this via multimeter. Note the placement in relation to pins 3 and 5 of the NJM2058 op-amp. The markings correspond to the center and end connections, respectively (see schematics below). **CAUTION**: Solder directly to the surface mount resistors and DO NOT allow solder to flow onto the adjacent NJM2058 pins (see finished mod picture below).

These schematics show the six connections from the four 47k-ohm resistors (see below). Wires 2 & 5 make up the 'center' connections, whereas, wires 1 & 3 and 4 & 6 make up the 'end' pairs.

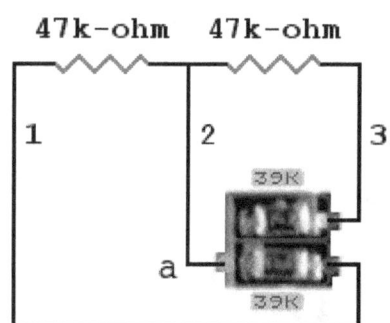
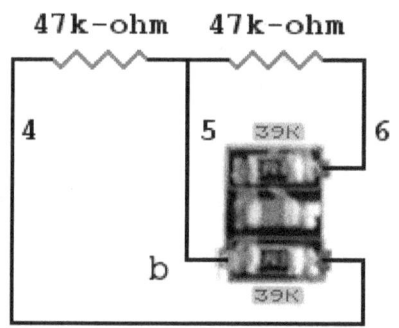

- The finished mod should look similar to the picture below.

- Check visually that no short exists. DO NOT leave any tools inside the unit!
- Check everything again, close the unit back up and connect all cables.

7. SURFACE MOUNT SOLDERING TIPS

Please read the tips located in my ECSS Volume Mod article.

8. CONCLUSION

Your R75 will now have significantly increased audio fidelity. ICOM created the R75 primarily as a SSB communications receiver. Dr. Rado's Sync-AM and AM-AGC mods transformed the R75 into a real performer on AM (SAM). This mod further enhances world-band program and music listening by increasing high frequency audio response.

9. CREDITS

I wish to thank Ken for his determination, his knowledge, his soldering skill, and most importantly his friendship. Without his help there would be no HiFi mod.

1.4 Phil's Fidelity Increase Mod
Application Notes
©2002

1. FIDELITY ANALYSIS

Spectral analysis was performed on my fidelity modified R75 (see graph below). The vertical height represents strength (at each subsequent frequency) with solid base corresponding to ~30 dB and ragged top to ~5 dB. Using pseudo dual 15-kHz filters yielded ~7250 Hz of bandwidth (see 'A' below). Using 15-kHz and 6-kHz filters yielded ~5750 Hz of bandwidth (see 'C' below). Using dual 2.4-kHz filters yielded ~2750 Hz of bandwidth (see 'D' below). Using dual 2.4-kHz filters detuned by 1.2 kHz (as per Cookbook) yielded ~4000 Hz of bandwidth (see 'E' below). Note that –6 dB attenuation points are used to specify filters; however, a strong signal attenuated to a fourth of its original level is often still audible. The faint line at 7000 Hz may be an artifact.

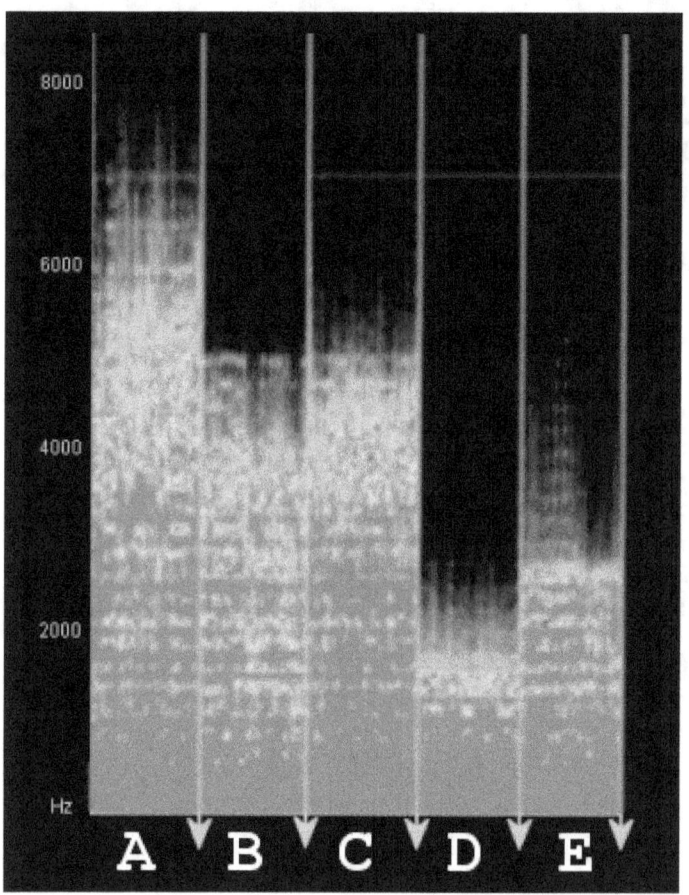

2. DSP FUNCTIONING

The second plot (see 'B' above) is the result of using pseudo dual 15-kHz filters with DSP NR engaged at its lowest setting of one. This plot [when compared with 'A' above] suggests that the DSP is able to function up to 5000 Hz. Frequencies (signal and noise) beyond this point are lost by engaging the DSP (this can be very useful). Using a carrier as a heterodyne, further testing revealed that the ANF was effective up to ~5000 Hz as well.

3. RESISTOR OPTIONS

Adding four 47k-ohm resistors in parallel theoretically yields ~5500 Hz of fidelity. In reality 5500 Hz represents the –6 dB drop point of each active two-pole low-pass filter. Combined these filters provide –80 dB per decade drop-off meaning that a 5500 Hz signal is attenuated to one-hundred-millionth of its original size. Another viable option is adding four 100k-ohm resistors in parallel for a theoretical yield of ~4200 Hz of fidelity.

4. SWITCH OPTION

Although I feel a switch is totally unnecessary it is still an option for those wanting to change between stock and fidelity-enhanced sound modes. A 4PST (4-pole single-toggle) switch and its placement on lines 1, 3, 4, and 6 are shown below.

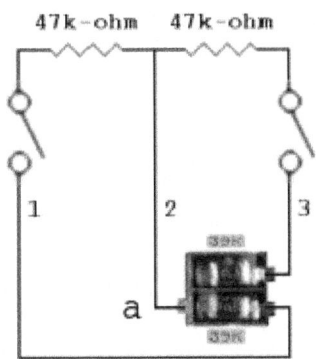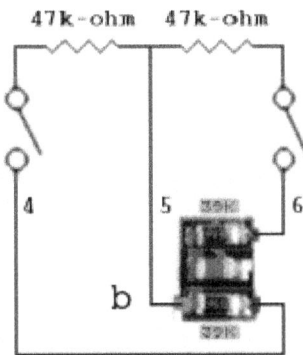

5. COMMENTS

The fidelity mod will significantly increase high-frequency audio response in the ICOM R75. This modification is meant to enhance the Sync-AM and AM-AGC mods.

1.5 Phil's Cheap AM Filter Mod
©2002

WARNING: DO NOT perform this mod without some type of **eye protection**.
DISCLAIMER: The author is not responsible for any damage resulting from this mod.

1. ABSTRACT

Add a 3.0 kHz AM filter for only $19; no soldering inside the R75 required!

2. INTRODUCTION

This modification will add a high-performance 15-element ceramic Murata filter (CFS455J) in the (455 kHz IF) optional filter slot inside the R75. This filter has the following characteristics: 3 kHz bandwidth at –6 dB, 9 kHz bandwidth at –80 dB, 8 dB maximum insertion loss, and 2000 ohms input/output impedance. This filter is excellent on AM and is especially nice for MW-tuning in that it does not allow adjacent station bleeding yet sounds better than the narrow 2.4 kHz filters. This filter's mellow sound is apparent more by its shape and ultimate rejection than by its –6 dB bandwidth specification of 3 kHz. Drawbacks: this filter cannot be used with SAM and leaks (flutter noticed) on SSB. Schematics of this filter and the ICOM filter slot are shown below.

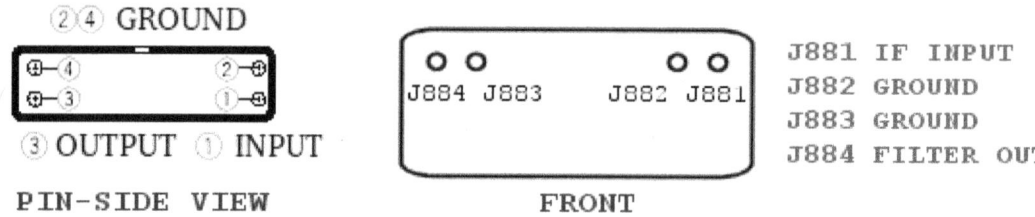

3. WARNING

Make sure to unplug the unit before beginning and wear eye protection. Disconnect all cables ESPECIALLY the power supply, and avoid static discharge by grounding yourself.

4. MATERIALS

To perform this modification you will need the following items:

- A Murata filter (see below).
- A resistor (any value, see methods section below).
- A tie-wrap, some wire, electrical tape or heat shrink tubing.
- A soldering iron, holder, sponge, and solder.
- Safety glasses or goggles (please use while soldering).

The Murata filter model CFS455J is available at Surplus Sales of Nebraska for $7. They have a $10 minimum so I ordered two and paid $5 shipping; technically two radios can be done for $19. Surplus Sales' telephone number is 1-402-346-2939. Their e-mail address is grinnell@surplussales.com. Website filter link is shown below.

http://www.surplussales.com/Filters/Filters-4.html

5. METHODS

☻ Solder short (~3.5 cm) pieces of wire to each of the four filter pin ends. Cover these connections using heat shrink or electrical tape (pink arrow in second picture). Terminate the ends of the four pieces of wire with a short (~1 cm) segment of wire cut from a standard resistor (see below). Note: use a clean (new) resistor or lightly sand an old one for good electrical contact.

☻ Take off the top cover by removing eight screws (see user manual). Tighten a tie-wrap around the filter (orange arrow) and then slip the R75's metal filter carrier (blue arrow) between the filter and the tie-wrap until the filter is seated securely. Refer to the diagram below while attaching the filter as follows: The filter's center protruding 'pin' should be facing the back of the receiver (located below at the green arrow's tip). The front left pin goes into the hole marked J884 (refer to the numbers), the rear left pin goes into the hole marked J883, the front right pin goes into the hole marked J881, and the rear right pin goes into the hole marked J882. Make sure the pins are seated tightly in their holes (the hole narrows further down for a snug fit).

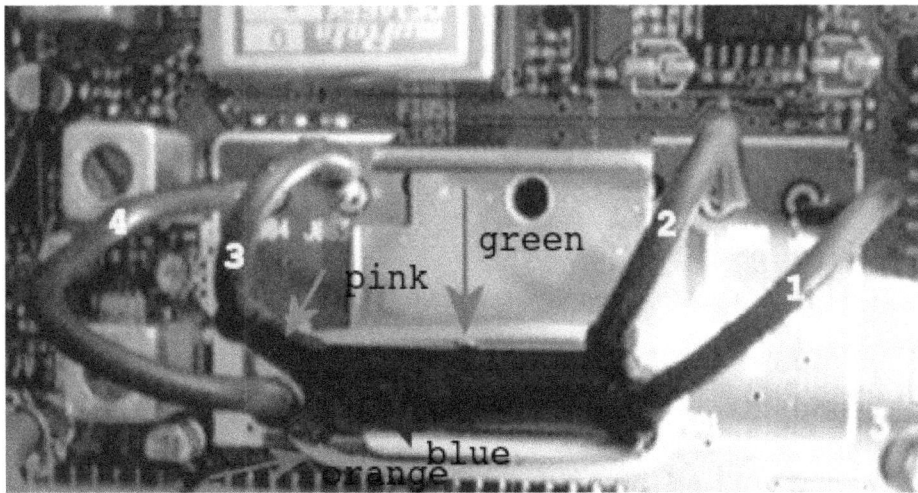

☻ Check the connections, close the unit back up, and connect all cables.

6. USAGE

Enter the filter setup mode by depressing 'FIL' 2 seconds. Using 'UP' find 'ExP oFF' and rotate the tuning knob to the right until 'Exp oN' is displayed. Using 'UP' find 'oP2 No' and rotate the tuning knob to the right until 'oP2 257' is displayed. Now you can set a filter value of '3.3 kHz' (please refer to the Cookbook) for usage. The 'N' stands for 'narrow' and the 'W' stands for 'wide'. Press 'FIL' a final time to exit the filter setup mode.

I now use the following (9 MHz & 455 kHz) filter settings on AM:

- AM wide = 15 kHz & 6 kHz
- AM normal = 15 kHz & 3.3 kHz
- AM narrow = 2.4 kHz & 2.4 kHz

7. CONCLUSION

Your R75 now has another AM filter! This filter mod is inexpensive, effective, and requires no soldering inside the R75.

8. CREDITS

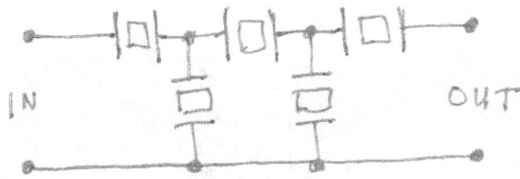

I would like to thank my good friend Ken for his help. This is as much his mod as it is mine. Thank you Enrique for catching my wiring error and Craig for your suggestions. The old version of this document showed the filter input and output switched. My belief is that the Murata 15-element ceramic ladder filter is symmetric (similar to the diagram above) and as such I did not notice the error.

1.6 PhilMod ECSS Volume Increase
©2002

WARNING: Performing this mod will void your warranty and could **destroy** your radio.
WARNING: DO NOT perform this mod without some type of **eye protection**.
CAUTION: This mod takes soldering skill so please practice beforehand.
DISCLAIMER: The author is not responsible for any damage resulting from this mod.

DANGER: This mod is new as of April 2002 and is being presented here for evaluation purposes. Please wait for any feedback from our electronics wizards, Ken, and I before attempting this mod!

1. ABSTRACT

Increase SSB (ECSS) volume by soldering a resistor in parallel with R1102. The value of the added resistor ranges from ~300 (high gain) to ~1300 ohms (low gain). R1102 is the gain resistor on the SSB mixer.

2. INTRODUCTION

This modification will increase the audio volume of SSB reception and ECSS (the tuning of AM stations as if they were SSB). This mod coupled with the R75's stability and tuning accuracy allows ECSS to become a viable listening alternative to AM. ECSS reduces fading, reduces distortion, and allows selection of the sideband (either LSB or USB) with the least interference.

3. THEORY

Below is a block diagram of the FM, SSB, and AM lines that enter the audio frequency select chip (IC1201). The lines in bold are where SSB amplification ideally should occur.

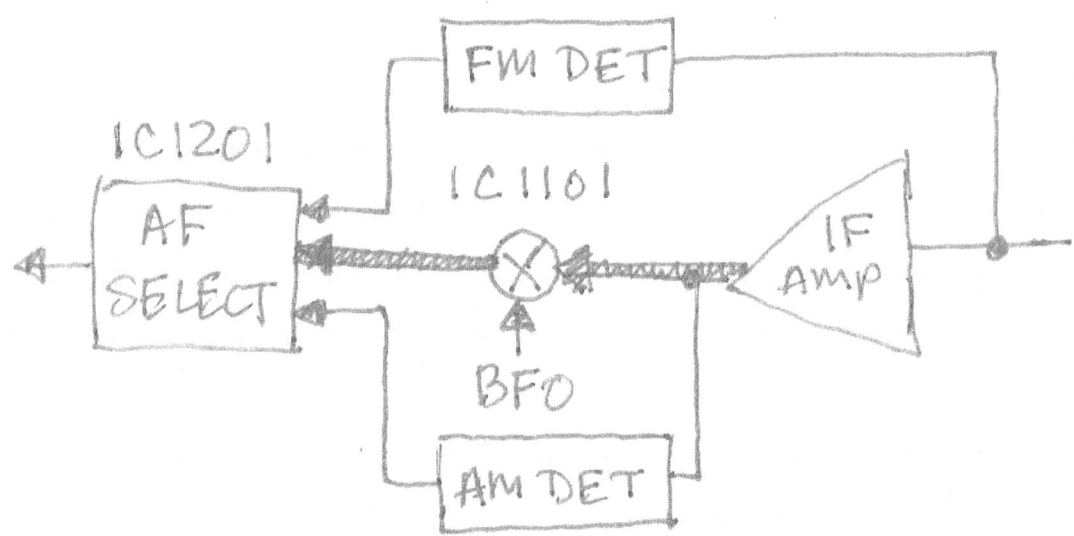

Below is a schematic of the bold area above: shown is resistor R1102 that limits the gain of the signal coming from the SSB mixer (MC1496D). Placing resistance in parallel with the 1.5k-ohm resistor (R1102) lowers total resistance. This lowered total resistance directly increases the SSB mixer's gain.

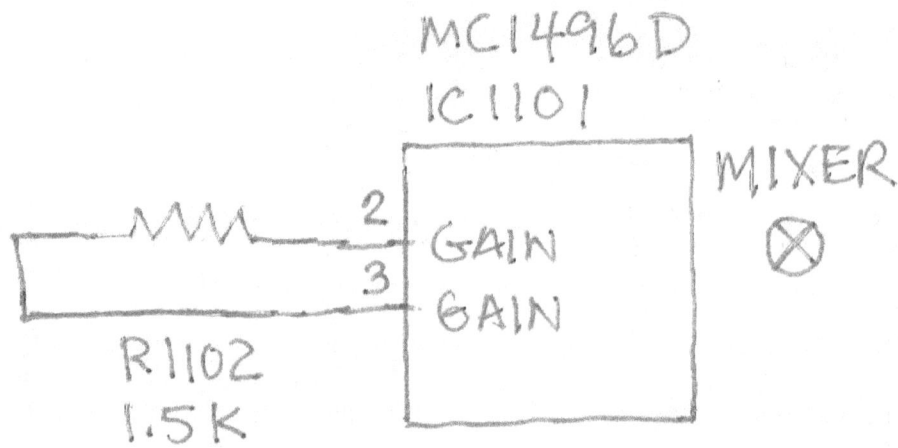

The schematic below shows (arrow) how to attach resistance 'in parallel' with R1102.

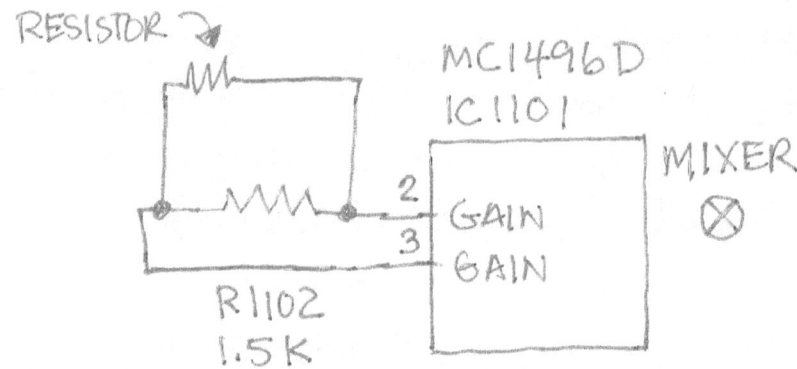

Other amplification ideas were initially considered. Resistor R1201 was lowered with inadequate results and schematics were created for usage of an operational amplifier as well as for a mode-sensing AGC bleed-off resistor. The two, complex, latter circuits proved unnecessary.

I believe the volume difference between AM and ECSS exists because of the AM signal's effect on the main AGC as well as the fact that the Motorola AM stereo chip has an onboard AGC circuit capable of over 1200% signal gain. This leaves the ECSS sound level weak by comparison. Not enough ECSS signal making it to the pre-amplification chip often necessitates the AF gain being turned up very high. This mod equalizes the volume by increasing SSB mixer gain and allowing the main AGC circuit to balance out the difference. This is not an ideal fix but it is as good as I could come up with and maintain simplicity.

Audio spectral analysis revealed that the stock R75 has a 10 dB difference between AM and ECSS while using the same filter. Under normal circumstances this becomes 20 dB because SSB filter bandwidths are half the width of their equivalent AM counterparts. Using this mod AM and ECSS volume is more similar although the larger bandwidth on AM brings in more volume with music. Larger ECSS filters should minimize this effect.

4. WARNINGS

Please read the warnings at the top of this document! Unplug the unit before beginning and wear eye protection. Soldering skills are a must: doing this mod could destroy your radio.

5. SIDE EFFECTS

This modification may cause SSB (ECSS) mode volume to be fairly loud with the AGC OFF. This can be adjusted for by reducing the RF gain a bit. This mod may necessitate reducing the RF gain till the s-meter reads in the 30 dB range for strong signals. The VFO may also need minor adjustment for some signals. This mod can easily be reversed by unplugging the unit and simply cutting and terminating the two added wires or by desoldering them.

6. PREPARATIONS

Before you begin: read this article in its entirety, gather all materials, disconnect all cables (ESPECIALLY the power supply) and avoid static discharge by grounding yourself. Be careful that the soldering iron does not burn any cables!

7. RESISTANCE OPTIONS

Deciding on a resistor value is a personal choice. Ken likes a value of 1300 ohms (low gain); whereas, I like a value of 400 ohms (high gain). A middle value such as 850-ohms may be a good initial value to try. One way to make 'trying' easier is to get a 300-ohm resistor and put it in series with a 1000-ohm pot, as show below. For the variable resistor use is the 1k-ohm 15-turn PC-mount trimmer pot: ex. Radio Shack (#271-342). You will need to connect the center contact and one of the ends in the circuit. If taped to the R75's wall behind the display with the screw facing upward, it can be adjusted with the case open a crack. On my R75 there is only a 1000-ohm pot, as I do not believe shorting pin 2 and 3 will cause any problem other than pre-amp overload; however, I recommend the more cautious and larger gain range setup mentioned above. Make certain the screwdriver does not fall into the unit while adjusting as it could cause a short.

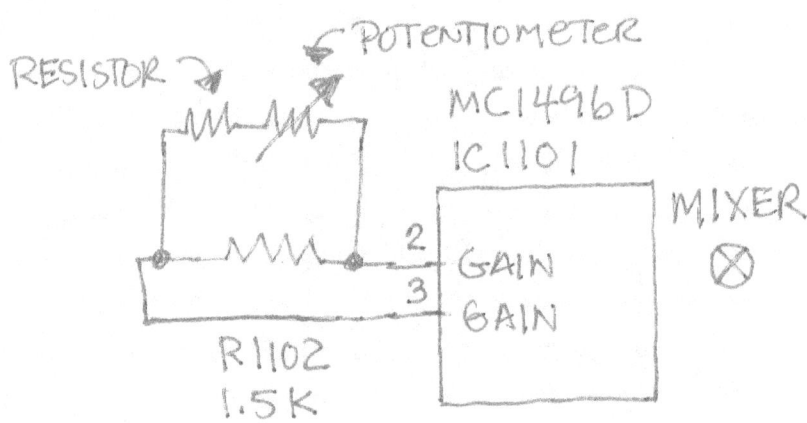

8. MATERIALS

To perform this modification you will need the following items:

- A resistor (and possibly a potentiometer, see 'Resistance Options' above).
- Thin insulated wire (ex. wire wrap).
- Electrical tape.
- A low-wattage needle-tipped soldering iron (holder, sponge, solder).
- A philips head screwdriver.
- Safety glasses or goggles (please use while soldering).

9. METHODS

- Take off the top cover by removing eight screws (see user manual for details).
- Locate the general region where R1102 is located from the picture below (square).

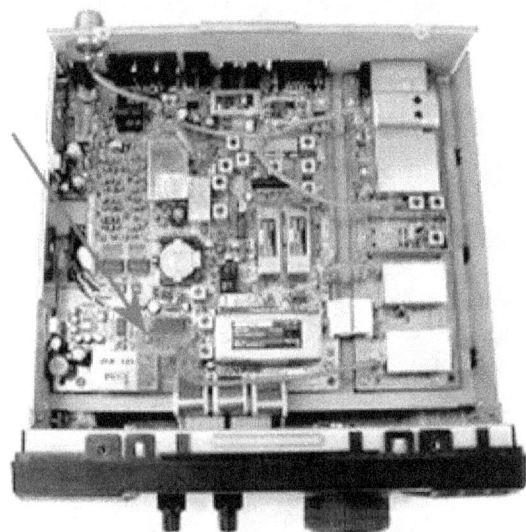

- Locate resistor R1102 (1.5k ohms) as pointed to by the white arrow in the picture below. Using a magnifier you will note the resistor has the marking '152' that stands for 1500 ohms. You can check this via a multimeter. Note that pins 2 and 3 of the mixer (MC1496D; underlined below) attach directly to R1102 as per the schematic above.

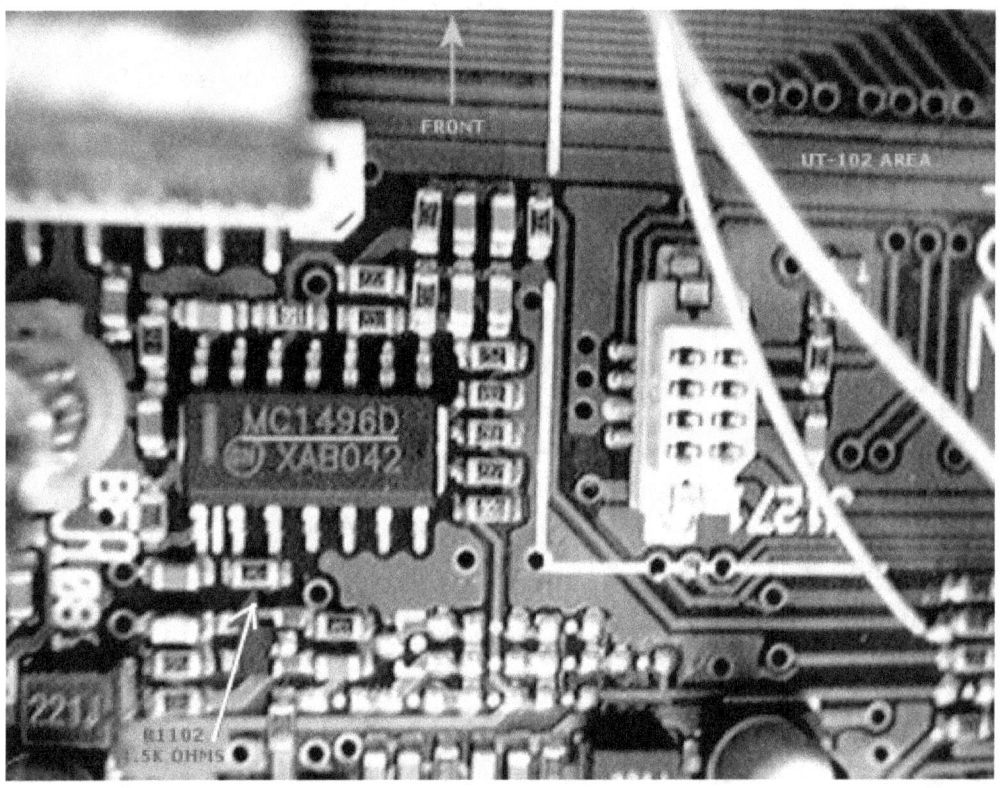

- Clip short the ends of the resistor (and potentiometer if used), solder both ends to some wire, use either heat shrink or electrical tape (or both) to cover all exposed metal, and tape this to the metal wall 'behind' the display area. If a potentiometer is used make sure the screw faces upward to allow for easy adjustment (see arrow below).

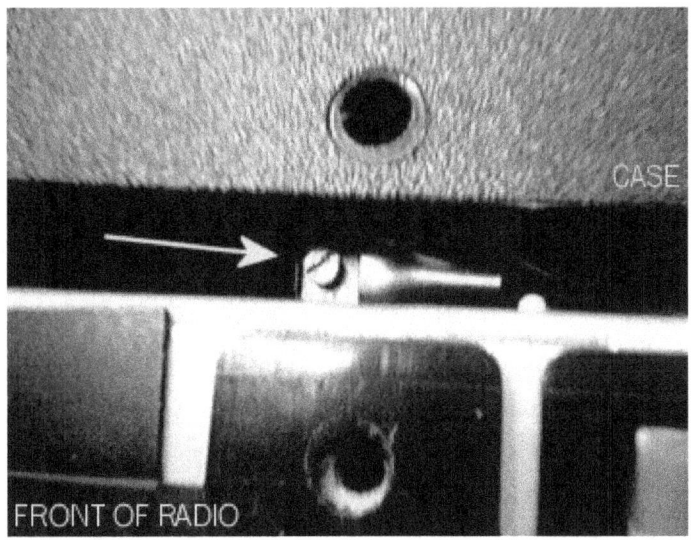

🪰   Solder one end of the wire to one side of R1102. Solder the other end of the wire to the other side of R1102. It should look similar to the picture below. Note the long wires coming from the added resistance.

🪰   Check visually that no short exists across resistor R1102. You can also connect a multimeter: if there is almost no resistance then a short exists. Make sure not to leave any tools inside the unit!

🪰   Check everything again, close the unit back up, connect all cables, and enjoy.

10. SURFACE MOUNT SOLDERING TIPS

Here is the technique I used: clip some wire wrap so only 2 mm is exposed, apply solder to the iron, dip the wire into the solder until only very little solder gets on the wire, **wipe the excess off the iron**, touch the wire to the end of the surface mount component (a resistor here), heat the solder with a needle tip iron for a few seconds until there is good flow and you are done. If by some chance you use too much solder and it flows into another component DO NOT PANIC: take desoldering braid, add some solder to it as this will aid flow, put over-top of the mess, heat the solder you added to the braid, and press until you see the mess flow up into the braid. **DO NOT OVERHEAT**: if you are not done any step within a few seconds, stop, wait for everything to cool, and then go back. Lastly, practice first: please DO NOT make this mod your first attempt at soldering surface mount. I practiced by making ~30 connections on an old cellular phone PCB and I 'was' an experienced solderer.

11. CONCLUSION

Your R75 will now be louder in SSB mode and during ECSS reception of AM signals. I hope this mod makes your R75 experience all the more enjoyable. Ken and I really like this mod and I find myself using ECSS all the time now.

12. CREDITS

I would like to thank my good friend Ken for being brave enough to be the first to test the various incarnations of this mod. Without his help there would be no mod. I also thank Dr. Rado for his guidance. And to the several people whose names cannot be mentioned here (you know who you are): thanks! Lastly I want to thank the R75 community: it remains one of the big reasons this receiver is so great!

1.7 Pete's MW Attenuation Removal Mod
©2002

WARNING: Performing this mod will void your warranty and could **destroy** your radio.
WARNING: DO NOT perform this mod without some type of **eye protection**.
CAUTION: This mod takes soldering skill so please practice beforehand.
DISCLAIMER: The author is not responsible for any damage resulting from this mod.

1. ABSTRACT

This is a description of Pete Gianakopoulos' MW attenuator bypass mod.

2. INTRODUCTION

The stock R75 is attenuated by 10 dB on LW and MW. Pete found that shorting the MW attenuator pads (see arrow in picture above) resulted in only a 2 to 3 dB increase in sensitivity. Pete subsequently created a working MW attenuator bypass mod that he describes below:

"It seems that 100 Ohm current limiting resistors are used as the series Rs with the switching diodes in the front end. In an earlier post, I mentioned that the shunt Rs in the MW attenuator were providing the return path for the switching diode in the front end. I calculated the equivalent resistance of this network to be 139 Ohms; that is the reason that I specified the 2.2 mH choke in series with the 110-Ohm resistor. Anyway, I did change the 10-Ohm series resistor that ICOM has in their unit to the 100-Ohm unit that is used in all of the other filter ranges; this eliminates the need for the extra resistor that is in series with the 2.2 mH choke. I didn't like the fact that the 2.2 mH choke is self-resonant at around 900 kHz, so I switched that value to a 1 mH Murata surface mount inductor. It may be because I eliminated some of the parasitics that were present with the 2.2 mH choke, but the sensitivity on the LW band improved. The 0.1 uV spec is now maintained down to 150 kHz, degrading to 0.2 uV at 100 kHz. This is because of the noise floor rise at 100 kHz and below, maybe due to the close-in phase noise of the synthesizer."

3. METHODS

Pete describes how to perform the modification as follows:

"To remove the MW attenuator, remove R171 and R173 (270 Ohms), and replace R172 (18 Ohms) with either a zero Ohm resistor short that connection point pad that is supposed to bypass the attenuator (it really does not, unless you remove R171 and R173). Next, change R174 (10 Ohms) to a 100-Ohm resistor. Finally, add a 1 mH surface mount choke across the point where R171 was connected. This is to provide the DC return path for the switching diode. You could use a leaded choke, but the main thing is to keep the leads as short as possible. On a final note, I must caution anybody that these parts are relatively small; we are talking about 0603 (.06 by .03 inch) parts. You really need to have some sort of magnifier or stereomicroscope or close up reading glasses to do this job. Also, you need a good soldering iron with a small tip, such as a Pace, Metcal, or Hayco soldering station to do the job. The reason for this is that you need to be able to concentrate the heat in a very small area. If anybody has any questions, feel free to give me a shout, and I will walk you through the job."

a ▨ short the connection pads
b ▨ remove R172
c ▨ replace R171 with a 1 mH surface mount choke
d ▨ remove R173
e ▨ replace R174 with a 100-Ohm resistor

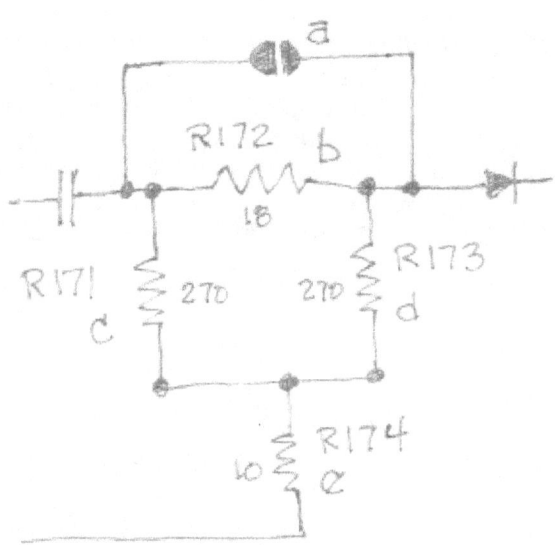

4. CREDITS

Pete Gianakopoulos of Chicago, Illinois created this modification.

1.8 Display Frequency Calibration
©2003

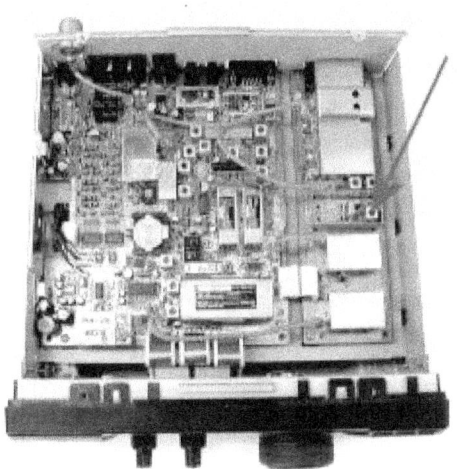

CAUTION: This procedure has not yet been tested.
DISCLAIMER: The author is not responsible for any damage resulting from this procedure.

1. ABSTRACT

The following procedure corrects frequency display error.

2. METHODS

- Turn the receiver off and remove the six case screws.
- Obtain a plastic "tuning" screwdriver.
- Locate L2 (right PCB, inside a metal fence) shown in the arrow above and close-up below.

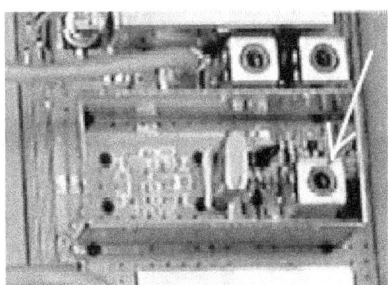

A. Turn the receiver on for one hour to allow warm-up.
B. Press "TS" until the triangle disappears.
C. Depress "TS" for 2 seconds until 1 Hz resolution is displayed.
D. Tune to 5000 kHz, 10000 kHz, 15000 kHz, or 20000 kHz.
E. Turn L2 until USB, LSB, and AM sound equal (white arrow above)
F. Turn the receiver off and replace the six case screws.

WARNING: Only use a plastic tuning screwdriver and DO NOT drop screws into the unit.

1.9 Voice Synthesizer Volume Adjustment
©2003

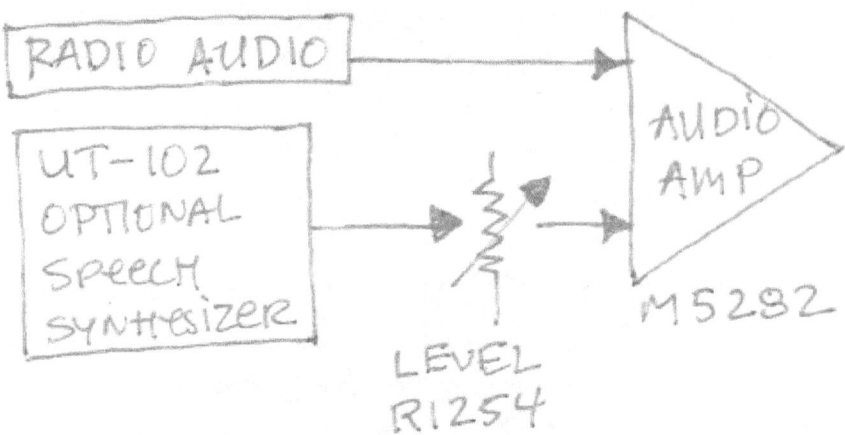

CAUTION: This procedure has not yet been tested. **USE A PLASTIC SCREWDRIVER**.
DISCLAIMER: The author is not responsible for any damage resulting from this procedure.

1. ABSTRACT

The following procedure alters the volume on the optional UT-102 voice synthesizer.

2. METHODS

- Turn the receiver off and remove the six case screws.
- Using a **plastic screwdriver** adjust R1254 (see circle below).
- Replace the six case screws.
  Notice the M5282 amplifier (circle) and R1254 resistor in the block and picture.

1.10 DrPhilSCAN Cookbook
Easy Tuning Mode in 2002
©2015

1. INTRODUCTION

PhilSCAN was software that: *controlled the operation of* and then *analyzed the sound of* the ICOM IC-R75. One version was written in PowerBASIC due to its small IDE and inline x86 assember.

2. IC-R75 RS232 CONTROL

The stock IC-R75 can be controlled via an RS232C cable. The R75 menu settings are as follows: *ADR = 5A, baud = HI, trn = on,* and *731 = off.* The code below, during WM_INITDIALOG, sets up the computer's COMM port. A MessageBox prompted for COM1, COM2, COM3, or COM4.

```
COMM OPEN "COM1" AS #2    'In actual code "COM1" is passed via a string.
COMM SET #2, BAUD     = 19200
COMM SET #2, BYTE     = 8
COMM SET #2, PARITY   = %FALSE
COMM SET #2, STOP     = 0
COMM SET #2, TXBUFFER = 0
COMM SET #2, RXBUFFER = 1024
```

The subroutine iCOMOut simplified sending data to the R75 by adding a header string and terminating character. During WM_INITDIALOG (dialog callback message) the R75 was turned on.

```
SUB iCOMOut(BYVAL commdata AS STRING)   'Form Packet & Send Data to R75
    COMM SEND #2, CHR$(&HFE,&HFE,&H5A,&HE0) + commdata + CHR$(&HFD)
END SUB


iCOMOut CHR$(&H18,&H01)   'R75 Power On
```

The subroutine freqUpdate changed the R75's frequency. The code below shows how the sent data was built. Shortwave frequencies were scanned from 2.3 MHz to 26.1 MHz, in 15 bands.

```
SUB freqUpdate (BYVAL freqNavigate AS LONG)    'R75 Change Frequency
    fx    = FORMAT$(freqNavigate, "0000000000")
    fBCD$ =           CHR$( (ASC(fx,9)-48)*16 + ASC(fx,10)-48)
    fBCD$ = fBCD$ + CHR$( (ASC(fx,7)-48)*16 + ASC(fx,8 )-48)
    fBCD$ = fBCD$ + CHR$( (ASC(fx,5)-48)*16 + ASC(fx,6 )-48)
    fBCD$ = fBCD$ + CHR$( (ASC(fx,3)-48)*16 + ASC(fx,4 )-48)
    fBCD$ = fBCD$ + CHR$( (ASC(fx,1)-48)*16 + ASC(fx,2 )-48)
    iCOMOut CHR$(&H05) + fBCD$   'RS232 Output
END SUB
```

Before a scan, four commands were sent to the R75. The frequency was decreased by 1200 Hz so that a carrier can be seen at 1200 Hz USB. The ANF must be off or it will delete the carrier. A 50 mS pause was used between iCOMOut calls. During WM_CLOSE the COMM port was closed.

```
iCOMOut CHR$(&H14,&H01,&H00)  'Volume Off      (&H16 Volume Up after scan)
iCOMOut CHR$(&H16,&H12,&H01)  'Super-Fast AGC  (&H03 Slow AGC  after scan)
iCOMOut CHR$(&H06,&H01)       'USB Mode        (&H11 SAM Mode  after scan)
iCOMOut CHR$(&H16,&H41,&H00)  'ANF Off         (Automatic Notch Filter Off)


COMM CLOSE #2   'At Program Exit
```

3. PC SOUND CARD ADC INPUT

A PC has an analog-to-digital converter (ADC) in its sound card input. RS232 commands steered the R75 to a frequency, the ADC gathered data, a Fourier transform (FFT) was run, and then values were analyzed. Below, the sound card was setup: 16-bit, 22.050-kHz, and PCM format.

```
GLOBAL TheAudio AS WAVEFORMATEX
TheAudio.wFormatTag       = %WAVE_FORMAT_PCM
TheAudio.nChannels        = 1
TheAudio.nSamplesPerSec   = 22050
TheAudio.wBitsPerSample   = 16
TheAudio.nBlockAlign      = (1 * 16) \ 8
TheAudio.nAvgBytesPerSec = ((1 * 16) \ 8) * 22050
TheAudio.cbSize           = 0
waveInOpen(BYVAL VARPTR(deviceHdl), %WAVE_MAPPER,
           BYVAL VARPTR(TheAudio), 0, 0, 0)
waveInStart(deviceHdl)
```

The code below was run in a loop until scanning ended (or the user aborted). Note that ADC data was sent to the RealInput() array via a pointer. The last three lines of code ran when the process ended.

```
STATIC TheWave AS WAVEHDR
TheWave.lpData          = VARPTR(RealInput(0))
TheWave.dwBufferLength = 1024
TheWave.dwFlags         = 0

waveInPrepareHeader(deviceHdl, BYVAL VARPTR(TheWave), LEN(TheWave))
waveInAddBuffer(deviceHdl, BYVAL VARPTR(TheWave), LEN(TheWave))

DO
LOOP UNTIL ((TheWave.dwFlags AND %WHDR_DONE) = %WHDR_DO)

waveInUnprepareHeader(deviceHdl, BYVAL VARPTR(TheWave), LEN(TheWave))
     CALL waveInReset(deviceHdl)   'Abort ADC input
CALL waveInClose(deviceHdl)   'Abort ADC input
```

4. FFT AND ANALYSIS

The next step was a software FFT. The FFT consists of an outer loop; wherein the block size starts at 2 and doubles to 1024 (the number of samples). This takes 10 iterations. Calculations include: 2 sine, 3 multiply, and 1 divide. The middle loop undergoes 10 iterations. And the inner loop, consisting of eight multiplications (etcetera), is done up to 1024 times. Real *integers* are input and complex values with real and imaginary *single-precision floating-point* parts are output.

The maximum and second maximum were looked for over a specific frequency range (near 1200 Hz). The absolute value of the FFT's real output was utilized. A flag was set if the maximum was centered. A call was then made to display a block whose color depended on the amplitude of that signal. Contrast could be user controlled via a scroll bar. The frequency was changed. And the process re-started. When done an array sort allowed the frequencies to be sorted by amplitude.

```
color = ABS(ADC_FFT_Amplitude(frequency) * 4) + contrast - 112
```

2.1 SSB Cookbook
Single Sideband for Portables
©2013

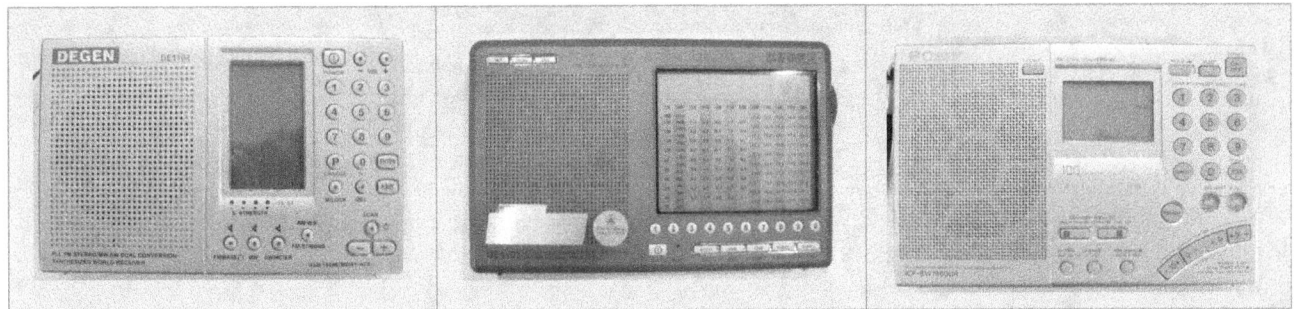

1. SSB PORTABLES

Many portables do not include SSB; including some I recommend for shortwave listening (PL-390, PL-380, and KA2100). Some previously recommended radios with superb SSB are now discontinued (Sat800 and E1). Some dual-conversion, PLL-synthesized portables do not have SSB (DE1105, KA11, KA1101, and PL-450). And others have poor SSB (DE1121, G4000A, ATS-505P, PL-600, and PT-80). The G6 Aviator has 10-Hz synthesizer steps but overloads. The Satellit 750 is big, costly, and has AGC troubles. The G6 Aviator and Satellit 750 have quality control issues. The ATS-909X has 40-Hz synthesizer steps but is large, costly, and has a poor knob. The G3 has overload and quality control issues. The PL-660 is good on SSB but was beat by the SW7600GR.

2. SSB PORTABLE PICKS

It might come as no surprise that the KA1102, DE1103, and SW7600GR are my SSB picks. These radios are double-conversion and PLL-synthesized; which makes for good: gain distribution, image rejection, and stability. Each of these radios is excellent for shortwave listening and has its schematic available online for study. The SW7600GR uses both a filter and phasing for selectivity. It should be noted that tabletops, such as Yaesu's FRG-100 or ICOM's IC-R75, have superior SSB.

Radio	RF	BW	Step	SSB	Cost	Size	Mem	Knob	Rech
KA1102	GOOD	2	1 kHz	YES	$69	22	133	NO	YES
DE1103	GOOD	2	1 kHz	YES	$72	31	268	YES	YES
SW7600GR	GOOD	1	1 kHz	YES	$145	45	100	NO	NO

RF (radio and audio performance), BW (bandwidths), Size (in cubic inches)
Mem (shortwave memories), Rech (recharging system and batteries)

3. HOW TO TEST SSB

Radio	Filters	Phasing	USB/LSB switch	Upper range	Lower range	Total range
KA1102	2	NO	NO	+4 kHz	-2 kHz	6 kHz
DE1103	2	NO	NO	+1 kHz	-1 kHz	2 kHz
SW7600GR	1	YES	YES	+1 kHz	-2 kHz	3 kHz

Single sideband (SSB) ranges were calculated by: 1) tuning the portable to an AM station [double sideband with carrier] (ex. 800 kHz) while in SSB mode, 2) de-tuning in 1-kHz steps (ex. 801 kHz for USB), and 3) turning the BFO wheel to hear if the station could be resolved (Donald Duck sound fixed). The DE1103's 2 kHz range made fine tuning easy. The KA1102's 6 kHz range allowed the wide and narrow filters to be used (added SSB fidelity). Each radio was good at SSB.

4. HAM RADIO BANDS

Band	Voice range in MHz	Size in kHz	Activity	Sideband
160M	1.800 - 2.000	200	LOW	LSB
80M	3.600 – 4.000	400	LOW	LSB
40M	7.125 – 7.300	175	MEDIUM	LSB
20M	14.150 – 14.350	200	HIGH	USB
17M	18.110 – 18.168	58	MEDIUM	USB
15M	21.200 – 21.450	250	MEDIUM	USB
12M	24.930 – 24.990	60	LOW	USB
10M	28.300 – 29.700	1400	MEDIUM	USB

The chart above reviews the amateur bands; where SSB communication can be heard. A good place to start listening to SSB ham audio is on the 20 meter band (14.150 to 14.350 MHz).

5. HOW TO USE A BFO

The SW7600GR's sideband switch makes for easy LSB/USB tuning. Most portables have only a BFO dial. Although more complicated, I have found an easy way to pre-tune the BFO for SSB. This is how to use a BFO: 1) tune to a strong AM station (ex. 3865 kHz), 2) engage SSB, 3) engage a narrow filter, 4) for USB up-tune by 1 kHz [about half the size of the narrow filter] (ex. 3866 kHz), 5) for LSB down-tune by 1 kHz (ex. 3864 kHz), 6) turn the BFO wheel until the voice clears up, 7) enter the frequency for one edge of the ham band in question, 8) tune using 1 kHz steps, and 9) make only **minor** changes to the BFO dial, if needed. This procedure works for most hams because they tend to park on 1 kHz boundaries. Use USB to listen to the 20M ham band.

6. DISCUSSION

The KA1102 is small and light. It has a vivid display, a blue backlight, and backlit buttons. Its large SSB range allows SSB using the narrow **or wide** filter. Unfortunately, this range, mixed with a thin BFO wheel, means it takes some finesse to accurately tune SSB. The KA1102 must be switched to **Page 9** to engage SSB. The radio is not as sensitive as the other two. Adding a small wire to its short antenna can help. Tuning through bands can be tedious due to the lack of a knob.

The DE1103 is sensitive and resists overloading when using the external antenna jack (it disengages the RF amplifier). It has a yellow backlight, and backlit buttons. It is inconvenient to press **VOL** to use the tuning dial as a volume control. Fortunately, the volume does not need to be changed often. And volume can be changed by keying a volume number, then pressing **VOL**.

The SW7600GR is sensitive, well built, and has an easy to deploy antenna. Included is a wind up antenna and premium carry case. Missing is an AC adapter ($20 aftermarket). There is no circuitry to charge batteries internally. The lack of a knob can make band scanning tiresome. The SW7600GR has big buttons. Ironically, the firmware makes the radio fairly unintuitive to use.

7. TOP SSB PORTABLE

If forced to choose one portable for SSB, it would be the DE1103. It is missing a volume knob but is a lot of radio for $72. I have heard many SSB transmissions using Degen's DE1103. The DE1103 is a good all-around portable for SSB/AM on MW/SW, as well as FM on the FM band.

2.2 DE1103 Cheat Sheet
Using your DE1103.
©2011

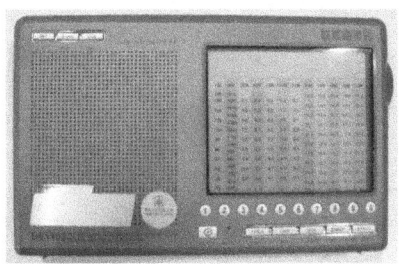

The DE1103 is a great portable radio for DX due to its sensitivity and overload immunity. The highlighted functions below are especially useful. Offset tune in AM and SSB mode to reduce adjacent interference and selective fading distortion. Offset by about half the filter bandwidth (1 to 3 kHz). Example: To tune 9980 kHz in AM mode use 9978 kHz for LSB and 9982 kHz for USB. To tune 9980 kHz in SSB mode use 9979 kHz for LSB and 9981 kHz for USB (then use SSB wheel).

Gray in the sequence column indicates a button must be depressed and held.
OFF indicates that the radio must be off.
BOLD indicates use of a knob or switch.

Function	Sequence	Comments
SET VOLUME	VOL **KNOB**	Stops after 2 seconds of inactivity.
SET VOLUME	<0 to 63> VOL	Easy volume control.
SET TIME	TIME **KNOB** TIME **KNOB** TIME	OFF Time is in 24 hour format.
VIEW TIME	TIME	Toggle. Shows time for 5 seconds.
SET SLEEP	SLEEP **KNOB**	OFF Hold SLEEP and turn knob.
SET SLEEP	<1 to 99> SLEEP	OFF Turns radio off after 1 to 99 minutes.
ON/OFF ALARM	M/F or STORE	OFF Toggles On/Off. M/F is 1. STORE is 2.
SET ALARM	M/F **KNOB** hour, minute, memory volume, duration (minutes)	OFF Press M/F or STORE after each choice. Use M/F for ALARM1 or STORE for ALARM2.
SSB (FM STEREO)	SSB	Red LED ON. Use **FINE** tune wheel (SSB). Offset tune: up for USB, down for LSB.
FILTER	**NAR/WIDE** SWITCH	News/NAR is 4 kHz. Music/WIDE is 6 kHz.
ATTENUATOR	**LOC/DX** SWITCH	LOC turns it ON. DX turns it OFF.
BAND	BAND- or BAND+	There are 12 band memories.
KNOB TUNE	**KNOB**	Tuning steps: 1-kHz on AM, 25-kHz FM.
DIRECT TUNE	<frequency> BAND+	BAND- for FM entries.
AUTO TUNE	BAND- or BAND+	Scan, pause 3 seconds, then continue.
MEM TUNE	M/F **KNOB**	Toggles On/Off. MEM on display.
STORE MEM	STORE **KNOB** STORE	Hold STORE and turn knob.
STORE MEM	<1 to 99> STORE STORE	Easy memory store.
RECALL MEM	<1 to 99> M/F	Easy memory recall.
DELETE MEM	STORE **KNOB** TIME	Hold STORE and turn knob.
DELETE ALL	SLEEP HOLD TIME RESET	Hold all three keys and do a RESET.
BACKLIGHT	**ON/OFF** SWITCH	Using an AC adapter, ON is continuous.
KEY LOCK	HOLD or HOLD	Press to turn ON. Hold to turn OFF.
CHARGE	VOL **KNOB** (hours)	OFF Hold VOL and turn knob.
RESET	**HOLE**	Use a paper clip to reset the CPU.

2.3 The DE1102 Mini-Cookbook
©2006

1. INTRODUCTION

The DE1102 is a portable combining digital tuning, PLL frequency synthesis (little drift), double conversion (reduces images), two bandwidths (adjusts tone and helps reduce adjacent interference), decent SSB (reduces selective fading distortion), 133+ SW memories, diminutive size, light weight, and low cost ($64). Each unit includes an AC adapter, three NiMH rechargeable batteries, an external wire antenna, stereo earphones, a manual, and a cloth pouch. I recommend purchase on eBay from a seller that includes the needed 220V to 110V AC converter.

Many portables fall short by comparison. To help with selective fading either SAM or SSB is necessary. The KA1101, ATS-606AP, SW40, and SW35 have no SSB while the YB-80, YB-400PE, G4000A, ATS-818ACS, and ATS-505P have poor SSB. The ATS-909 is large and costly. The SW07 is very costly. The SW7600GR features SAM and is an *excellent SWL choice* albeit larger and more expensive. The DE1103 is better for DX due to its sensitivity and overload immunity; however it is larger and SSB is not as good (newer DE1103 units have improved SSB).

2. TUNING TIPS

Manually tune in AM mode using the narrow bandwidth to insure stopping on the center frequency. Once there switch to the wide bandwidth for music (increased treble) or stations in the clear. With adjacent interference keep the narrow bandwidth and detune higher or lower by one or two kHz. This same procedure allows adjustment of tone (fidelity). For selective fading chose the narrow filter, switch to *Page 9*, engage SSB, detune by plus three or minus three kHz, and fine tune using the BFO wheel. The DE1102 is quite good on SSB for a low-cost portable. Tip: tune ECSS using wide until minimum warble then switch to narrow. The narrow filter will also improve sensitivity, lower noise, and decrease pumping. Filters work best when fed low power signals. Therefore, engage LOCAL (versus DX) and/or reduce antenna length as long as 3 or 4 of the red signal meter LEDs are still lit. Overly strong signals in AM mode or SSB mode will cause distortion.

For usage on the ham bands first tune to a high power AM (DSB) signal, chose the narrow filter, switch to *Page 9*, engage SSB, detune by minus two kHz for LSB or plus two kHz for USB, and fine tune using the BFO wheel. Then switch back to the ham band and as long as the amateur is broadcasting on a 1 kHz boundary they will become audible when stepping through the band. The displayed frequency will be off by 2 kHz (low on LSB and high on USB) from what a tabletop would show. For example: 3863 kHz on the DE1102 display would represent 3865 LSB on a tabletop; likewise, 14.250 MHz would represent 14.248 MHz USB.

On MW the unit selects the internal ferrite antenna: the radio must be rotated from front facing to side facing for optimal reception. On FM antenna height and direction are critical. FM has several options including mono or stereo, music or news (via headphones), and bass boost. The DE1102 backlight comes on automatically (photosensor) when buttons are pressed in the dark. Note: If the AC adapter is plugged in while backlighting is on it will remain on only if the sensor detects darkness (cover the LCD with your hand before and during attachment of the DC plug).

3. INITIAL SETUP AND CHARGING

Function	Sequence	Comments
TIME FORMAT	ENTER ENTER <1 or 2>	Remove battery for ~30 seconds.
MW STEPS	ENTER <1 or 2>	Remove battery for ~30 seconds.
CHARGING	**CHARGE** 7 ENTER	OFF Seven hours for full charge.

OFF indicates that the unit must be off for the function.
Note: Battery removal [MCU RESET] retains memories but not the clock or alarm.

4. TIME RELATED FUNCTIONS

Function	Sequence	Comments
SET CLOCK	ENTER ENTER <time>	OFF
VIEW CLOCK	EXIT	
SET ALARM	ENTER ENTER ENTER <time>	OFF BEEP outputs through speaker only.
VIEW ALARM	EXIT	OFF Alarm BEEP duration is 60 seconds.
CANCEL ALARM	ENTER ENTER ENTER EXIT	OFF
SET SLEEP	EXIT ENTER <01 to 99>	Sleep defaults to 99 minutes.
CANCEL SLEEP	EXIT ENTER EXIT	Sleep is on by default.
SET "ON-OFF" TIMER	ENTER <S1-S2-S3> <time> <FM-MW-SW> <frequency**> <VOLUME> [ENTER <01 to 99>]	OFF S1-S2-S3 are speaker symbols. ** ENTER confirms last preset as frequency. OFF time starting with ENTER is optional.
VIEW TIMER	<S1-S2-S3>	OFF
CANCEL TIMER	<S1-S2-S3> EXIT	OFF
ACTIVATE TIMER	<S1-S2-S3> ENTER	OFF

GRAY indicates that the button must be depressed and held.

5. MEMORY AND SCANNING

Function	Sequence	Comments
SET PAGE	P <0 to 9>	P0: **ATS**. Paging alters tuning steps... P1-P6: FM-50kHz, MW-9/10kHz, SW-5kHz P7: FM-10kHz, P8: MW-1kHz, P9: SW-1kHz
STORE MEMORY	M/LOCK <digits>	Use 1- for numbers 10 through 19.
RECALL MEMORY	<digits>	Use 1- for numbers 10 through 19.
DELETE MEMORY	DEL DEL	P0: remaining memories compressed for **ATS**.
MEMORY TUNE	SCAN+-	Press SCAN until display reads MEM.
MANUAL TUNE	SCAN+-	Press SCAN until display reads STEP.
AUTO SCAN	SCAN+-	Press SCAN until display reads STEP. P1-P6.
MANUAL SCAN	P 9 SCAN+-	Press SCAN until display reads STEP. Hold SCAN+- down continuously. Scan speed automatically slows on signals.
AUTO TUNE SCAN (ATS)	P 0 SCAN+-	Press SCAN until display reads STEP. Enter frequency and memory to start from.
COPY PAGE 0	P 0 P 0 <1 to 9>	Band must match for P7-P9.
DELETE PAGE	P <0 to 9> DEL DEL	Hold DEL till the page number blinks.

SCAN+- are the (-) and (+) keys below the SCAN button.

6. MISCELLANEOUS

Function	Sequence	Comments
FREQUENCY ENTRY	ENTER <digits>	EXIT clears digits or backs out. ENTER for quick input of trailing zeros.
SW METER BANDS	SW	49M-41M-31M-25M-22M-19M-16M
FM BASS BOOST	FM	For headphones as is MUSIC-NEWS.
KEYPAD LOCK	M/LOCK	Toggles on and off.

7. ETON E1 COMPARISON

The Eton E1 is an excellent SWL radio. However there are attributes that the DE1102 has over the E1. The DE1102 is 13% of the cost ($64), 10% of the size (21 cubic inches), and 19% of the weight (12.3 ounces with batteries). The DE1102 comes with a carry loop, protective pouch, earphones, rechargeable batteries, 13 foot external antenna, MW ferrite rod, and 70 MHz to 76 MHz FM coverage. The DE1102 has convenient battery access and charges batteries within the unit; a photosensor driven blue backlit display and keyboard; no display hash or ghosting; FM bass boost, an optional 12-hour clock format, three event timers, automatic tune scanning (ATS), a commonly available antenna connector, and an attenuator. Note that DE1102 display contrast is best when the unit is laid face (speaker) upward.

SWL involves reception of relatively high powered DSB stations. In side by side testing the DE1102 using its included 13 foot wire antenna heard almost all the signals present on the E1 using its whip. This included both SW and ham signals. The DE1102 had the advantage on MW because its internal ferrite rod could null out competing stations on the same frequency.

Although the DE1102 has no SAM its SSB capability is sufficient to handle both one-sided interference and reduce selective fading distortion. The E1 SAM is much easier to use than the manual BFO of the DE1102. Adjusting the BFO takes finesse and every now and then voices start sounding mechanical due to drift. During selective fading the E1 SAM sounded better when listening to music. However the DE1102 is quite capable when tuned to hams, news, or talk radio programming. Music sounds good on the DE1102 in AM mode when there is no selective fading.

It was surprising to hear programs sounding better on SSB using the DE1102 than on SAM using the E1. That said: the E1 is outstanding on SSB. SSB often sounds better than SAM during selective fading because the resultant audio is disturbed less during carrier dropouts. The DE1102 experiences negligible SSB warble when detuned as per the "Tuning Tips" section above.

Obviously the E1 is the technologically superior radio. However I find myself using my DE1102 more than my E1 because it is so portable: 21 cubic inches, 12.3 ounces, chargeable batteries, and convenient carry loop. *The downside is that SSB drift mandates frequent retuning.*

8. CONCLUSION

The DE1102 makes an ideal ultra-portable for SWL. Using the DE1102 does necessitate learning some tuning tricks and retuning the BFO when using ECSS (tuning DSB signals as SSB to reduce selective fading distortion). The incredible $64 price tag reflects the Chinese devaluation of their Yuan. Although having no SAM, the *Degen* DE1102 offers the portability of the $355 SW07. For portable SWL I *highly recommend* this remarkable little radio.

2.4 KA2100: Superior by Design
©2008

1. INTRODUCTION

I highly recommend the DE1102, DE1103, and KA2100 portables. Performance comparison revealed the KA2100 as champion on MW, FM, and SW. Examination of each radio's internals and schematics helped illuminate how *Redsun Electronics* achieved this feat. The KA2100, with its lack of SSB and unrefined software, could easily end up hastily dismissed. Beneath the silver and black exterior resides an old-school PCB and solid RF engineering. The KA2100 makes use of mechanical controls. It is possible that the KA3100 will end up our next king of low-priced portables.

Radio	RF	BW	Step	SSB	Cost	Size	Mem	Knob	NiMH
DE1102	GOOD	2	1 kHz	YES	$64	22	133	NO	YES
DE1103	GOOD	2	1 kHz	YES	$64	31	268	FAIR	YES
KA2100	GOOD	2	1 kHz	$33	$100	229	30	GOOD	$18

2. DESIGN COMPARISON

The MW section of each radio consists of a ferrite rod antenna followed by an RF amplifier. The KA2100 uses Twin-Coil technology invented by *Chris Justice*. Each end of the ferrite contains a pick-up coil that feeds into a combining transformer. MW tuning in the KA2100 is done via MCU control and is transparent to the user. Stand-alone Twin-Coil antennas cost $100.00 at C. Crane.

The FM section of each radio is diode switch, enters a TA7358 FM front-end chip, goes into two 10.7 MHz filters, and finishes in the tuning chip. The KA2100 has a separate FM demodulator, the LA3335. The KA2100 alone has a tuned BPF and post-amplification uses dual tuning elements.

Radio	MW Twin-Coil	MW Tuned	FM RF	FM Tuned	FM Demod
DE1102	NO	NO	LC	SINGLE	TA8132
DE1103	SOME UNITS	NO	BPF	SINGLE	TA2057
KA2100	YES	YES	TUNED	DUAL	LA3335

The AM (and MW) section of each radio consists of a BPF, LPF, attenuator, dual JFET mixer, BPF, single JFET mixer, two IF filters, AM/FM tuning chip, and an MCU controlled PLL, the LC72137 (tuning both AM and FM local oscillators). The KA2100 alone has input protection diodes, RF gain control, and five metal shielded areas on the PCB. Note the DE1102's lower first IF of 10.700 MHz.

Radio	Diode Protect	RF Gain	1ST IF MHz	2ND IF kHz	Tuner Chip	Boxed Shield	SSB
DE1102	NO	NO	10.700	450	TA8132	2	YES
DE1103	NO	NO	55.845	450	TA2057	3	YES
KA2100	YES	YES	55.845	455	LA1260	5	NO

The audio section on the physically larger KA2100 is beefier and drives a five inch speaker. The KA2100 has separate bass and treble control knobs. Unlike in many radios, these tone circuits actually work. The KA2100 utilizes a TDA2822 Dual Power Amplifier run in bridge configuration. It is buffered using a whopping 8800 microfarads of capacitance and can output five watts of power.

Radio	Bass/Treble	Audio Amp	Buffering	Power
DE1102	2 settings	LM4811; LM4862	220 µF; 470 µF	210 mW; 400 mW
DE1103	2 settings	CXA1622	220 µF	??? mW
KA2100	2 knobs	TDA2822	8800 µF	5000 mW

3. E1 VERSUS KA2100

The $500 E1 is superior to the KA2100 in numerous ways. However, the KA2100 also has a laundry list of better features, it has a: folding handle; heavy knob with dimple; backlit keys; crisp display; internal transformer; easy battery access; firm battery holder; 'AA' or 'D' battery power; NiMH charger; firm antenna; separate MW antenna jack; tuned Twin-Coil ferrite rod; two antenna adapter plugs; metal speaker grill; RF gain; attenuator; IF output; 5-kHz SW plus 9/10-kHz MW knob steps; 12 hour time format; and no safety recall. Ironically the technically lesser KA2100 has better audio than the E1 which could be due to the sloppier filters and better bass-treble controls. The KA2100 is definitively better on MW due to the antenna; while slightly more sensitive on SW.

The KA2100 is not without its negatives, the biggest is its memory recalling system. There are only 10 hard-to-recall memories per 10-MHz SW segment. Memories should be accessible via the tuning knob! The KA2100 has no keypad and that would be fine with ±0.5 MHz up/down slew buttons (equivalent to two full knob turns). Other negatives include two noisy band-change beeps, delayed button response, SSB via a $33 external 4 'AA' powered unit, and use of 32Ω headphones.

Note that the DE1102 and DE1103, which cost two thirds as much, include earphones, long wire antenna, NiMH batteries, wrist loop, cloth pouch; in a smaller size and with integrated SSB. One plus the KA2100 has on these radios is its 5-kHz knob tuning steps: for fun manual scanning.

The outward appearance of the Eton S350DL is akin to the KA2100. However the S350DL is a second-rate, drifty, analog-tuned, single-conversion radio utilizing a digital frequency counter.

4. USE EARPHONES and TURN KEY LIGHT OFF

Using 15,000 mAh Duracell alkaline "D" batteries the KA2100 can be powered indefinitely (8.5 years) with the radio off. With speaker power, batteries will last ~54 hours. The big surprise is that using earphones (32Ω needed) the KA2100 will last ~202 hours! Stretching battery life is easy: use earphones, turn the key light off (switch), and turn the backlight off (via LIGHT button).

KA2100 Power Consumption						
Quiescent	MW/SW	FM	RF-gain	Backlight	Keylight	Earphones
0.2 mA	277 mA	-4 mA	+2 mA	+12 mA	+35 mA	74 mA
75000 hr	54 hr	55 hr	53 hr	52 hr	48 hr	202 hr

Add (+) or subtract (-) from 277 mA; earphones value is for MW/SW.

5. DISCUSSION

In my opinion the KA2100 is an instant SW classic. The engineers at *Redsun Electronics* in China deserve a round of applause. They went the extra mile with each section of the radio [MW, FM, SW]: designing for excellence. The radio uses the same chip (TDA2822) and a similar speaker (8Ω; 5W) to that used in the Eton E1. Except for the two loose-fitted band-switch knobs the radio is extremely well built; as is the AC cord. It will be interesting to see what a KA3100 brings.

6. KA2100 QUICK REFERENCE

This guide only includes non-obvious functions. With the internal speaker select FM MONO; use FM STEREO with headphones. Press RESET following changing the MW steps from 9/10 kHz.

Function	Sequence	Comments
BACKLIGHT	LIGHT	Continuous backlighting. Note the display flash.
TURN OFF TIME	POWER	Pressing lowers automatic turn off time. Note: "on" means radio stays continuously on.
SEARCH	UP OR DOWN	DOWN Search stops on active stations.
SAVE MEMORY	MEM {UP|DOWN} MEM	Press UP or DOWN to select the memory slot.
RECALL MEMORY	PRESET	Pressing PRESET cycles through memories.
CHARGE BATTERY	PRESET	OFF Charging auto stops. Note: battery icon will flash.
LOCAL WORLD	W/L	OFF Change between World or Local display time.
TIME SET	MEM {UP|DOWN} <KNOB> MEM	OFF UP and DOWN cycles through time. OFF UP selects hours; DOWN selects minutes.
12/24 FORMAT	UP	OFF Note: AM/PM displayed in 12 hour format.
SET TIMER	TIMERA {UP|DOWN} <KNOB> TIMERA	OFF Same for TIMERB. OFF UP selects hours; DOWN selects minutes.
VIEW TIMER	TIMERA	OFF Same for TIMERB.
ACTIVATE TIMER	TIMERA ALARM	OFF Pressing ALARM cycles BUZZER, RADIO, and OFF.
ALARM OFF	POWER	Stops the alarm (3 minute buzzer or 60 minute sound).
SNOOZE	SNOOZE	Five minute snooze works four total times.
BUTTON BEEP	SNOOZE	Toggles button beep sound on and off.
QUICK TUNE	Q.TUNE	Steps: MW 100 kHz; FM 1000 kHz; SW by band.

GRAY indicates that the button must be pressed and held.
OFF indicates that the radio must be off for the function.
WARNING: *Only attempt to charge rechargeable batteries.*

2.5 FRG-100 SWL Tips
Using the Yaesu FRG-100.
©2007

1. INTRODUCTION

Although not recommended in my guides the Yaesu FRG-100 is great for SWL[1]. The FRG costs $600 new and ~$380 used. Strong points include the analog s-meter, bright display, and memory knob. Sensitivity is excellent and the FRG is good at DX. The tips below pertain to SWL.

2. USE SIDEBAND-SELECTED AM

I discussed using AM mode tricks to reduce selective fading distortion[2]. The FRG is ideal for these tricks because of its good audio (via headphones or external speaker), nice 4-kHz filter, and great slow AGC with a perfect 0.9 µV threshold. Engage the 4-kHz filter (AM/N), slow AGC (button out), and offset tune by 2.0 kHz upward for USB or 2.0 kHz downward for LSB. Using this AM trick the FRG is not hindered by lack of a SAM detector. During carrier dropouts the audio momentarily but minimally distorts and then returns to low-distortion, envelope detection. Detection on the FRG in AM mode consists of a silicon Schottky barrier diode (1SS198) followed by a ~7.2 kHz RC LPF.

This works best on new (1994+) units (so-called FRG-100B) where a LF-H2S filter replaces the CFW455I. The new tighter 4.5-kHz filter has superior shape (1:1.71) and ultimate rejection (70 dB). Offset tuning by 2.0 kHz simulates usage of a high fidelity ~8.5 kHz filter. The FRG has no keypad but the FAST tuning rate can be set to 5.0 kHz steps via: SET+FAST; dial in 5.0; SET. Using this setup one dial revolution covers 5 MHz and normal tuning is 0.1 kHz for offset tuning. An indoor tuned loop antenna can be used to null out local noise and is ideal for urban locations.

3. USE PRECISION ECSS

The FRG is good at SSB with its 10 Hz tuning steps and ±10 ppm stability. ECSS (tuning DSBc signals as SSB) is often superior to SAM during heavy selective fading[2]. The 6-kHz filter can be switched in briefly to zero ECSS: SET+SEL; dial in 6.0; tune till flutter stops; SSB (aborts filter change). The FRG allows audio shaping through a SSB offset adjustment: SET+MEM/CLEAR; dial in 453.40 for LSB and 456.60 for USB; SET. ECSS for SWL is rarely needed with the FRG.

[1] Phil's SW Radio Buying Guide
[2] Tuning Tricks Challenge SAM

2.6 Eton E1 SWL Tips
Using the Eton E1.
©2008

1. INTRODUCTION

For SWL I recommended the Eton E1 over other portables and tabletops[1]. Its $500 price buys Drake-engineered circuitry mated with consumer grade packaging, buttons, and knobs. The E1 is good for DX provided the antenna does not cause overload. The tips below pertain to SWL.

2. TURN DX MODE OFF

The DX mode turns on the +10 dB post-BPF RF-amplifier. For SWL the DX mode should be OFF because: 1) weak S2 (0.40 µV) signals can still be heard; 2) the s-meter is only accurate with DX mode off; 3) 5-kHz DR and IP3 limitations (worse than Eton quotes) are less pronounced; and 4) unknown to many, the AGC threshold is better. When DX mode is ON the threshold is no longer ideal. For SWL it is vital to use the slow AGC (3 second release; or "auto" mode) to reduce fading related distortion. An indoor tuned loop can null out local noise and prevent IP3 related problems.

3. USE ENHANCED SSB

The E1 is great on SSB with its 10 Hz tuning steps, "Enhanced SSB" (audio phasing; 30 dB of opposite sideband attenuation), high stability (better than quoted ±10 ppm), and PBT. The E1 SAM is great for music; however, SSB works very well on voice programs. I discussed how ECSS (tuning DSBc signals as SSB) can be superior to SAM during heavy selective fading[2]. Many do not know that SSB on the E1 offers ~70% less audio distortion than SAM mode. Note: turn "Enhanced SSB" off via menu during frequency alignments (screw is inside the leftmost rear ventilation slit).

4. USE THE 7-kHz FILTER WITH PBT

The 7-kHz filter is wider in bandwidth and tighter in shape than quoted. This is perfect for usage with the ±2 kHz PBT. Moving the filter away from the adjacent rejected sideband increases fidelity; moving towards decreases fidelity. The 7-kHz filter and PBT can simulate AM bandwidths of ~3 kHz to ~10 kHz (audio filtration limited). The 4-kHz filter simulates from ~1 kHz to ~8 kHz.

---

[1] Phil's SW Radio Buying Guide
[2] Tuning Tricks Challenge SAM

2.7 Eton E1 Flaw List
Design flaws of the Eton E1.
VERSION 1 ©2006

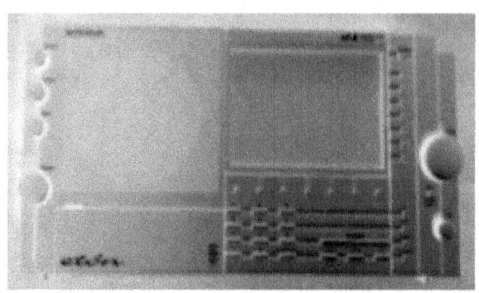

The Eton E1 is an excellent portable receiver but it is not without its failings. The tables below list some E1 imperfections that buyers should know about before purchase. I like the E1 and recommend the radio for SWL; however, it is best to learn about deficiencies up front. Quality control remains an unknown factor although it is likely satisfactory. It should be noted that *Passport 2006*, a superb publication, mistakenly stated that the 1200 country memories "***have a specific country permanently burned onto the top line of each page***". Country names *can be altered* by pressing and holding the COUNTRY button for three seconds.

MISSING
NO HANDLE MAKES CARRY INCONVENIENT
PAL TO SO-239 ANTENNA ADAPTER NOT INCLUDED
PROTECTIVE COVER NOT INCLUDED
EARPHONES OR HEADPHONES NOT INCLUDED
XM REQUIRES AN EXTERNAL ANTENNA MODULE
NON-MECHANICAL S-METER

DISPLAY
DISPLAY CONTRAST COULD BE BETTER
CONTRAST CONTROL IS IN BATTERY BAY
DISPLAY HARD TO SEE AT AN ANGLE
SOME HORIZONTAL DISPLAY GHOSTING
DISPLAY EMITS SOME RF RADIATION
NO PHOTOSENSOR FOR BACKLIGHT

POWER
RADIO CANNOT RECHARGE BATTERIES
BATTERY ACCESS IS INCONVENIENT
BATTERIES ROCK INSIDE RADIO WHEN MOVED
XM MODE DRAINS BATTERIES QUICK
ANTENNA JACK BECOMES HOT WITH AC USAGE
AC ADAPTER PRODUCES HEADPHONE HUM

PERFORMANCE
POOR 5 kHz DYNAMIC RANGE EVEN WITHOUT PREAMP
POOR 5 kHz DYNAMIC RANGE AND IP3 USING PREAMP
NO MW FERRITE ROD TO NULL COMPETING STATIONS
SPEAKER VIBRATES RADIO ON SOME LOW FREQUENCIES
NO ATTENUATOR
NO RF GAIN CONTROL
NO NOISE BLANKER, NOTCH, DSP NOISE REDUCTION

COMPARISONS
MORE ROOM FILLING AND VIVID SOUND ON SAT800
BETTER 20 kHz DYNAMIC RANGE ON SAT800
BETTER 5 kHz DYNAMIC RANGE ON 2010 AND SAT800 AND SW77
BETTER SYNC DISTORTION LEVEL ON SAT800 AND SW77
BETTER MW RECEPTION ON 2010 AND SAT800 AND SW77
BETTER PHASE NOISE ON SW77
BETTER BLOCKING ON SAT800

FEATURES
NO DIGITAL RADIO MONDIALE (DRM)
NO DIGITAL RADIO (HD)
NO RDS (STATION IDENTIFICATION)

SOFTWARE
NO 5 kHz SW OR 9/10 kHz MW TUNING KNOB STEP
SEEK SCANS OFF THE END OF EACH SW BAND
SSB FREQUENCY OFF AND CANNOT BE ZEROED IN SOFTWARE
NO AUTOMATIC TUNE SCANNING TO POPULATE MEMORIES
NO MENU FUNCTION TO SORT MEMORIES
TWO BUTTONS ARE NEEDED TO ENGAGE SSB FROM SAM MODE
HITTING XM WITH NO XM ANTENNA CAUSES 10+ SECOND PAUSE
MINOR PBT SOFTWARE GLITCH
NO 12-HOUR TIME FORMAT
NO RS-232 CONTROL ALTHOUGH LIKELY CAN BE IMPLEMENTED

ERGONOMICS
LIGHTWEIGHT TUNING KNOB
EASY TO ACCIDENTALY TOUCH KNOB AND TUNE OFF STATION
BUTTONS ARE TINY
NO TUNING KNOB FINGER INDENT
NO RUBBER GRIP ON TUNING KNOB
PBT LACKS MID-POSITION DETENT

MECHANICAL
TUNING KNOB USES MECHANICAL NOT OPTICAL ENCODER
ANTENNA CAN FALL DUE TO LOOSE SWIVEL ACTION
SMALL PLASTIC CIRCULAR CUTOUT IN BACK CAN BREAK OFF
NO HIGH IMPEDANCE SW ANTENNA HOOKUP
FM AND SW SHARE SAME EXTERNAL ANTENNA HOOKUP
HINGED REAR PANEL IS FLIMSY
STAND NEEDED FOR OPTIMAL USAGE

2.8 E1XM versus Sat800
Comparing radios with SAM.
©2005

1. FIRST IMPRESSIONS

The E1XM comes with a manual, CD-ROM, and 3-pronged thin-wired AC adapter. The unit is small, solidly built, and has a slightly rubber-like surface. The jacks, knobs, and buttons are all well laid out. Buttons are tiny but well spaced. The external antenna jack becomes hot with usage. The tuning knob moves easier than the other knobs but has no flywheel effect. *The Sat800 has a more substantial tuning knob* but the E1XM has better tuning step options. *The E1XM has no high impedance antenna input, no separate FM antenna input, and no carry handle.* It would have been nice if the E1XM included a carry pouch, headphones, and a PAL to SO-239 antenna adapter.

2. DISPLAY

The large 5.7" E1XM display (240 by 320 pixels) has three brightness levels (dim, medium, bright) and turns off after 10 seconds on batteries to conserve power. The unit can show a battery condition bar graph. The frequency is displayed in large rounded numbers in either kHz or MHz. In memory mode the contents of ten memories are displayed simultaneously. The time is always shown and becomes enlarged when the unit is off. There is a hinged rear panel for angled viewing. The user interface is extraordinary due to the informative display and numerous menu buttons. Keypad entry beeps can be turned off. *The display shows slight horizontal ghosting. Some of the indicators do appear larger on the Sat800 display.*

3. PORTABILITY

Size	Dimensions	Volume*	Weight	Antenna
E1XM	13.0" x 7.5" x 2.5"	244 ci	4.2 lbs	39.25"
Sat800	20.5" x 9.4" x 8.0"	1542 ci	14.2 lbs	56.75"

Volume is in cubic inches.

Power	Current* (lamp)	Hours** (lamp)	Minimum	Batteries
E1XM	210 mA (260 mA)	71.4 (57.7)	3.6 Volts	4 "D"
Sat800	510 mA (830 mA)	29.4 (18.1)	5.7 Volts	6 "D"

For the E1XM subtract 35 mA for FM or add 140 mA for XM.
**Duracell 15,000 mAH "D" alkaline batteries and 0.25 Watt audio.*

The Sat800 is over three times the weight and over six times the volume of the E1XM. The E1XM has twice the battery life; triple with continuous display lighting. The front mounted E1XM battery bay exposes a display contrast control and microprocessor reset button.

4. CLOCK

The E1XM has two 24-hour clocks (local and GMT) and two event timers (alarms). The clocks can be automatically set while the unit is off via NIST frequencies (WWV or WWVH) by using stored daylight savings and local time offsets. There is a 1 to 99 minute sleep timer and variable (5, 10, 20, or 30 minute) snooze for alarms. *The Sat800 has two clocks and two timers.*

5. MEMORIES

The E1XM has 500 alphanumerically namable memories and 1200 country (any label, not just country names, may be assigned to each group of 10) memories. Up to fourteen characters (A-Z, 0-9, plus, minus, period, comma, space, backslash) can be selected for labels via the tuning knob. The frequency, AGC, bandwidth, SAM, and SSB settings are all stored. Memories can be stepped through via the tuning knob either individually or in pages of 10; or via paging keys (again by 10). Memories are non-volatile. Note: bands can also be stepped through using the tuning knob. *The Sat800 has only 70 memories.*

6. AUDIO

	External Audio Output (9V)	▲ AGC per 90 dB ▲RF
E1XM	3 Watt	2 dB
Sat800	1 Watt	6 dB

Both radios have 4" internal speakers. The E1XM can output 3 Watts of audio and accepts an auxiliary line input that goes through the bass, treble, and volume sections. The E1XM bass and treble knobs have center indents. *The Sat800 speaker gives more room filling sound.*

7. EXTENDED COVERAGE

	Extended Frequency Ranges
E1XM	76-90 MHz (FM) 2.3325-2.3450 GHz (XM)
Sat800	118-137 MHz (VHF AIR)

Both radios cover 100 kHz to 30,000 kHz (LW/MW/SW) as well as 87 MHz to 108 MHz (FM). *The Sat800 includes VHF AIR* while the E1XM boasts XM satellite radio.

8. FM

FM	Sensitivity @ 20 dB S/N	Image Rejection	Finest Tuning Step
E1XM	4.00 µV (1.50 µV)	55 dB	20 kHz
Sat800	4.00 µV	50 dB	100 kHz

Number in parenthesis is using the E1XM "DX" switch.

FM reception is similar on the E1XM and Sat800. However, the E1XM has greater sensitivity, better image rejection, and finer tuning steps.

9. MW

The E1XM lacks a directional ferrite MW antenna. The radio is still quite capable on MW. Serious portable MW DX would likely be done without an internal ferrite: instead opting for a 34" square, air-core, air-capacitor tuned loop (see NRC or IRCA) or Quantum QX loop. Ironically *the Sat800 contains a small ferrite rod* but the radio is too large to easily turn and null stations. Sadly the $70 BCL-2000 has a larger ferrite than either of these $500 radios.

10. SW

Sensitivity	SSB 10 dB S/N	AM/SAM 10 dB S/N
E1XM	0.50 µV (0.25 µV)	4.00 µV (2.00 µV)
Sat800	0.50 µV	2.00 µV

Numbers in parenthesis are using the E1XM "DX" switch.
Note: S1 and S2 represent 0.20 µV and 0.40 µV respectively.

The "DX" switch which slightly reduces battery life is necessary to boost E1XM AM/SAM sensitivity levels to those of the Sat800. Filter selectivity appears equal: the wide E1XM -6 dB filter rating of 7 kHz is similar to what was measured on the Sat800.

SW	Image Rejection	Finest Tuning Step	5 kHz IP3
E1XM	70 dB	10 Hz	-20 dBm
Sat800	60 dB	50 Hz	-20 dBm

Image rejection is better on the E1XM and tuning steps are finer at 10 Hz. The E1XM tuning knob steps on SW are 10 Hz, 100 Hz, and 1 kHz. *There is no 5 kHz tuning knob step on the E1XM.* The E1XM appears to have the same IP3 as the Sat800; however, to get equal AM/SAM mode sensitivity requires usage of the +10 dB post BPF amplifier which will degrade IP3 to -30 dBM. Fortunately the E1XM is quite sensitive without the DX switch: this is good because using the amplifier increases the likelihood of first mixer overload. Hopefully *Passport 2006* will publish other specifications such as synthesizer phase noise, dynamic range, blocking, ultimate rejection, and SAM mode distortion.

11. BLOCK DIAGRAM

	First IF	Post BPF	Post 2nd mixer	Synthesis
E1XM	45.000 MHz	+10 dB amp ("DX")	amp-filter-amp	2 DDS & 1 PLL
Sat800	55.845 MHz	-20 dB attenuator	filter-amp-amp	3 PLL

The E1XM uses a different first IF frequency and instead of an attenuator has a +10 dB selectable RF amplifier ("DX" switch controlled) following the band pass filters. The order of the 2nd mixer's IF filter and one of its amps is reversed. E1XM frequency synthesis utilizes Direct Digital Synthesis (DDS). *The Sat800 uses Phase Locked Loop (PLL) synthesis.*

12. SEEK FUNCTION

The E1XM has an excellent seek function that works in both VFO and memory modes. Seek makes it easy to find new stations or active memory stations. The S-meter (S1 through S9+60 dB) is calibrated and used by seek to stop on signals above the threshold set by the squelch knob. The E1XM digital S-meter is so good that it rivals the analog Sat800 S-meter. The LW and SW bands are scanned in 5 kHz steps; MW in 9 or 10 kHz (selectable) steps; and FM in 200 kHz steps. Seek on SW covers 105 kHz per second; consequently, the largest band, 19 Meters, takes under 7 seconds to seek. Squelch muting can be turned off so sound is heard during seek.

There is also a tagged memory scan (T-Scan) function which operates in two ways: memories above squelch can be stopped on for as long as the carrier is above threshold or stopped on for only five seconds.

The E1XM squelch allows for quiet VFO operation: frequencies below threshold (ex. stations gone off air) are muted. The squelch value is seen as segments (there are 20 bars) below the multi-segmented S-meter. *The Sat800 squelch only works on the VHF AIR band. Searching for stations on the Sat800 using the 100 Hz tuning knob steps or by stepping through the band is fairly tedious.*

13. SAM

The E1XM SAM has a ±1 kHz lock range as well as USB, LSB, and DSB settings. This mated with three IF filters (7.0/4.0/2.3 kHz), a good AGC (slow/fast/auto), tone controls (bass/treble), and a large speaker allow for excellent SW listening. The E1XM AGC "auto" setting selects slow except while tuning. The E1XM and Sat800 SAM units are comparable although DSB on the E1XM can come in handy. *The Sat800 exhibits slight hum from the power supply.*

14. PBT and ENHANCED SSB

The E1XM has a ±2 kHz PBT. Under AM/SAM the PBT functions similar to detuning. The E1XM offers 30 dB more opposite sideband rejection than the IF filter alone in SSB mode. This is accomplished via the audio phasing networks whose primary function is selection of a sideband in SAM mode. Using this "Enhanced SSB" with the PBT allows simulation of filters smaller than 2.3 kHz. The audio output is narrowed by using LSB and positive PBT or USB and negative PBT. *The Sat800 does not have PBT.*

15. SSB

Good SSB reception (ham or utility) is possible on the E1XM by using "Enhanced SSB" with the PBT. The PBT can often attenuate adjacent noise. Morse code reception is aided by using these two tools as well. *The Sat800 is not as capable on SSB* as the E1XM.

16. ECSS: TUNING DSB AS SSB

ECSS is important for DX due to SSB sensitivity and the lack of the need for lock. The fine 10 Hz E1XM tuning steps allow for good ECSS reception. *The Sat800 often experiences warble.* The difference is more noticeable on music. Occasionally ECSS on the E1XM sounded better than SAM on the Sat800. This occurred during tough fading where the Sat800 SAM sounds loud and filled with treble. ECSS does not pump much on fades. The PBT also proved useful on ECSS.

17. THE NEW KING

Radio	Cost	Memory	SAM	Filters	Steps	PBT	Sensitivity	5 kHz IP3
E1XM	$500	1700	Drake	3	10 Hz	YES	0.25 µV	-20 dBm
SW77	$470	162	Sony	2	50 Hz	NO	0.16 µV	-37 dBm
ICF2010	$350	32	Sony	2	100 Hz	NO	0.15 µV	-21 dBm

The ICF2010, SW77, and E1XM are all capable SW listeners due to their SAM units. A cursory sensitivity test of the ICF2010 or SW77 against the E1XM may cause one to prematurely dismiss the E1XM. The real advantage of the E1XM is on SSB (ECSS): this portable has the tuning steps of an R8B and the stability of an FRG-100. Note: E1XM size (volume) is 60% larger.

Like the R8B, the E1XM PBT mated with audio phasing works similar to the twin-PBT on the R75: each allows the simulation of multiple SSB filters. The R75 has the advantage of 1 Hz tuning steps and deeper attenuation afforded by the second IF filter. However, the E1XM audio phasing is so good close-in that even 10 Hz steps allow for reception without warble. *Audio phasing mated with PBT and 10 Hz tuning steps is a huge step forward for portables.*

Tune either Sony to a ham signal with adjacent noise (QRM). Using the E1XM select the 2.3 kHz IF filter and apply PBT. The audio can often be narrowed so that signal remains while the QRM is attenuated. The fine 10 Hz steps allow for natural sounding speech. If SAM reception is poor due to harsh fading simply try ECSS (SSB). Using the E1XM select the 7.0 kHz IF filter and then tune in 10 Hz steps until sound is natural. Some are hesitant to use ECSS due to prior experiences with portables lacking the tuning steps, stability, and adjacent attenuation to allow for good audio.

In many ways the E1XM even puts tabletops to shame. The E1XM SAM betters what is offered by the stock NRD-545, RX-350, R75, FRG-100, and R30. The E1XM speaker betters the ones found on the NRD-545, R8B, RX-350, R75, FRG-100 and R30. The E1XM has 1700 memories, an excellent display, a built-in antenna, and 76-108 MHz FM coverage. The E1XM makes a better first radio than the R75 unless there are plans for large antennas or extensive SSB listening.

Press the E1XM "POWER" button, press "MEMORY", set the SQUELCH to S7, press "SEEK". Then watch the unit quickly and silently search down the memories, highlighting the current one, until it stops on a memory whose signal is S8 or greater. The display shows simultaneously: time, S-meter, current frequency, meter band; the names and frequencies of 10 memories; as well as AGC, bandwidth, PBT, SAM, and SSB settings.

18. COMMENTS

The Drake engineered E1XM is destined to become a classic. It is a compact Sat800 with numerous extra features. Impressive E1XM attributes include its small size, build quality, clean audio, Drake designed SAM, three IF filters, good AGC, 10 Hz tuning steps, PBT, enhanced SSB, 1700 memories, seek function, huge display, well written manual, and excellent user interface. If the rest of the technical numbers pan out I can see *Passport 2006* giving the E1XM portable over 4 1/2 stars. *The E1XM was well worth the wait and the $500 price tag.* Drake deserves a medal.

E1XM Front

E1XM Back with Hinged Rear Panel

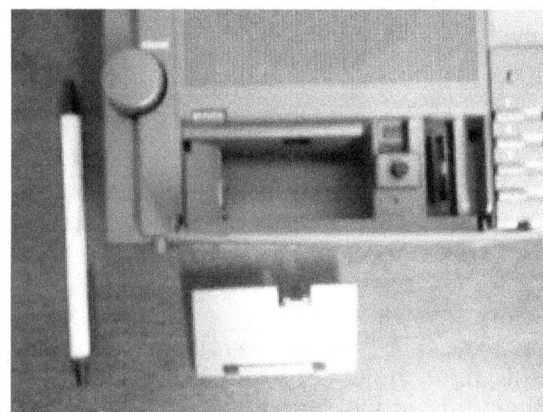

Battery Bay*

Manual, CD-ROM, and AC Adapter

black circle is display contract control: beneath it is the microprocessor reset hole.

E1XM and Sat800 Front

E1XM and Sat800 Top

E1XM and Sat800 Side

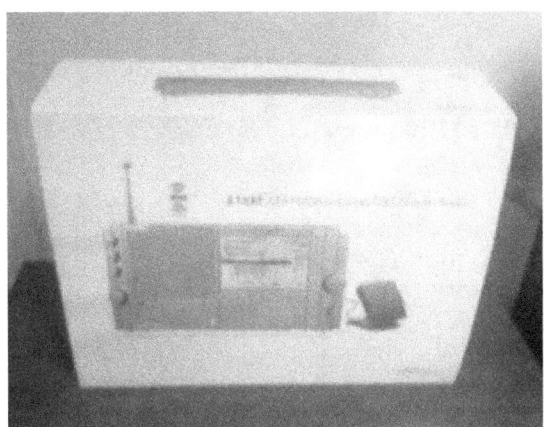

E1XM Box

2.9 E1 versus R75
Indoor Shootout From 2007
©2014

*The information below is **not** guaranteed to be error free.*

1. EQUIPMENT

This comparison is between a stock $500 Eton E1 portable and a modified $670 ICOM R75 tabletop. The R75 was altered equivalent to two *Kiwa Electronics* modifications costing $60 total: the *Synchronous Detector Upgrade* and the *High-Fidelity Audio Filter Upgrade*. The modifications were necessary to fix the SAM, overly fast AM/SAM AGC, and fidelity problems of the stock R75.

The MW antenna is a $200 *Quantum QX* loop acquired through *Radio Plus+*. The Quantum QX is a tunable loop with a tilting and rotating 7.5 inch ferrite head. This loop: has up to +40 dB of RF gain, is capable of 60+ dB nulls, and can operate portably on one 9V battery (using <8 mA).

The SW antenna is a $70 sixteen-inch *Torus Tuner* SW loop. The antenna was connected to an isolation 10:1 voltage balun to match the high impedance [500 ohm] loop to the low impedance [50 ohm] receiver inputs. Balun insertion loss was less than 1 dB. Loops work well in urban areas: they can be rotated to null out local noise. Tuned loops reduce first mixer energy.

Each antenna was selected, between the R75 and E1, using a *Daiwa CS-201* coaxial switch ($25 at Universal Radio). These switches are ideal for testing because of their smooth action. The switch provides over 50 dB of isolation, with an insertion loss of under 0.2 dB.

2. AUDIO

The audio of the E1 is better than the stock R75. Modifying and adding an external speaker to the R75 helps equalize the two. For SWL the E1 audio is superb. Meanwhile, for DX the crisp R75 audio and DSP noise reduction can make marginal signals sound better and scanning easier.

3. SAM

The modified R75 SAM nearly instantly gains lock. This helped when switching between the radios; however, on more severely fading signals the E1 SAM held on better. When the R75 SAM loses lock it can be set to seamlessly revert back to envelope detection. Whereas the E1 SAM can often be heard trying to regain lock. This reacquisition of lock on the E1 can take several seconds.

4. SSB

The E1 (phasing-PBT) and R75 (twin-PBT) both allow the simulation of multiple SSB filter bandwidths. The radios are both very capable on SSB. AM music broadcasts tuned via ECSS [SSB mode] sounded better on the E1. On SSB, the audio phasing unit helps the E1; whereas, the 1 Hz tuning steps help the R75. A superior design (to both) would be a radio with both audio phasing and 1 Hz tuning steps. With this setup and proper tuning, the SAM mode would be unnecessary.

5. MW

There were no signals that the modified R75 could identify that the E1 could not. This was expected. The Quantum QX provides a large amount of MW gain. MW band noise can be upwards of -135 dBm/Hz or S4 [2.0 µV; 2.3 kHz; 50 ohms]. A few very faint stations could be heard on the R75 that were not present on the E1; but none were identifiable. While testing, the knob tuning steps of 5-kHz and 10-kHz on the R75 came in handy; they aided manual signal hunting.

6. SW

The radios performed similarly on SW. Very faint signals were picked up better on the R75. These signals were below S1 [0.20 µV] in strength. On the R75 "PRE1" or "PRE2" were engaged; on the E1 "DX" was engaged. Using a passive tuned indoor loop only a small amount of DX was missed by the E1. For a portable the E1 was extremely capable during ham reception. This is due to the bandwidth narrowing phasing-PBT combination which works similar to the R75's twin-PBT.

7. DISCUSSION

The E1 and modified R75 performed similarly using indoor loops on both MW and SW. The E1 portable achieves tabletop-like performance when attached to indoor tuned loops, such as the Quantum QX or Torus Tuner. This is very good for the E1 considering that *Passport* called the R75 "*first-rate for unearthing tough utility and ham signals*". Large external antennas, beverages and random wires might favor the R75. The advantages of the E1 included its: big speaker, good SAM, phasing, large display, 1700 memories, ergonomics, and portability. R75 advantages included its: stability, 5/10-kHz and 1 Hz tuning steps, DSP noise reduction, display contrast, build quality, button feel, and beefy tuning knob (optical encoder). Ideally an E1 needs an angled viewing stand.

I highly recommend both the E1 and the R75. An ideal hybrid radio would have the R75's construction, 1-Hz steps, and ±1 ppm stability; and the E1's display, ergonomics, SAM, and audio phasing.

2.10 R8B versus R75
⬜2003

1. INTRODUCTION

The Drake R8B and ICOM R75 are popular shortwave receivers with loyal followings. These two excellent performers offer good ergonomics, numerous features, and analog [non-DSP IF] detection systems. This PDF will reveal information that is not readily found in advertisements or reviews.

2. PROFESSIONAL REVIEWS

Passport to World Band Radio 2002 Edition rated the R8B as "5 stars" stating: "The American made Drake R8B is the only non-professional receiver we have ever tested that gets everything right, where something important isn't missing or sputtering." *Passport* rated the R75 as "4.375 stars" stating: "The ICOM IC-R75 is a first-rate receiver for unearthing tough utility and ham signals, as well as world band signals received via manual 'ECSS' tuning."

World Radio Television Handbook 2000 Edition rated the R75 and R8B similar in: sensitivity, dynamic range, image rejection, RF intermodulation, IF performance, and audio quality. *WRTH* rated the R8B higher in built-in filter bandwidth choices whereas the R75 was rated higher for mechanical design, construction quality, and ergonomics. The R75 was awarded the *"Best Value Table-Top Receiver"* by *WRTH*. Since the award the R75's price has dropped 49% while the R8B's price has risen 22%.

Note: WRTH happened to review both receivers in their 2000 Edition.
Note: The stock R8B is known for its SAM-mode; the stock R75 for its SSB-mode.

3. THE MODIFIED R75

The stock ICOM R75 has three design flaws in the AM/SAM section of the receiver. These include a "too fast" AM AGC, a non-functional synchronous detector, and average fidelity. These flaws can be fixed for $80 at Kiwa Electronics with their "Synchronous Detector Upgrade" [$45] and "High-Fidelity Audio Filter" [$35] modifications. The modified R75 has yet to be professionally reviewed. Kiwa Electronics also offers a MW attenuator bypass mod to increase MW sensitivity [important to BCB DXers].

Beginners often do not realize that AM [double sideband with carrier] stations can be received using SSB. This is called ECSS tuning and it accounts for the good R75 reviews despite a flawed AM/SAM section. Both ECSS and sideband selectable synchronous detection allow selection of the half (sideband) of an AM [double sideband with carrier] station with the least interference. There is minimal "selective fading" distortion while using either synchronous detection or ECSS. The modified R75 can select a sideband during synchronous detection using a narrow filter; whereas, the R8B uses a method called phasing.

Modifications exist for the R8 series receivers that address: encoder shaft static, low headphone volume, broken front feet, synchronous detector heterodynes, synchronous detector noise, dynamic range, and image rejection. These modifications are not very popular; most Drake owners are not even aware they exist.

Note: The R8B and R75 both come with inadequate internal speakers. An external speaker is recommended.

4. THE DRAKE R8B

The Drake R8B costs $1470 and has the following features over a modified R75:

- Ease of usage (fewer tuning tricks to master)
- Five filter bandwidths included
- Large knobs and buttons
- Tone control
- Analog s-meter
- 1000 Memories
- Made in America

5. THE ICOM R75

The ICOM R75 costs $510 [or less, including DSP] and has the following features over a stock R8B:

- Sensitivity of 0.16 µV on SSB
- Selectivity: crystal SSB filter
- Dual-PBT: 400 Hz to 2400 Hz SSB bandwidths
- Two pre-amps
- 1 Hz tuning step resolution
- Low phase noise synthesizer
- High stability synthesizer
- Optional filter slots
- Extended frequency coverage to 60 MHz
- DSP noise reduction
- DSP automatic notch filter
- Complete computer (RS232) control
- Non-volatile EEPROM memories
- Backlit LCD (bright and sharp) display
- Molded faceplate
- Small size: mobile mountable and portable
- Quality buttons and knobs
- Solid and rubberized tuning dial

Note: Optional R75 filters are not usually necessary due to the twin-PBT.

6. SOME R8B DESIGN FLAWS

The R8B has the following flaws:

G. Power supply generates heat even when unit is off
H. Dial encoder failures (mechanical, not optical, encoders are used)
I. Synchronous detector noise and hiss
J. Audio hiss and bassy sound
K. Frequency drift
L. Display RFI
M. Multiple birdies
N. Wide shaped LC filter bandwidths
O. Unusual CW offsets
P. Buttons wobble on their centers
Q. Fussy passband offset
R. Alignments can be required

Note: Transformer heat can change electronic component tolerances, causing failure or the need for expensive alignments. An external power supply is recommended.

7. SOME R75 DESIGN FLAWS

The modified R75 has the following flaws:

- Display has no protective glass
- DC adapter can break at the PCB if excessive forced is used [solder fixable]

8. CONCLUSION

The $1470 R8B is arguably the best receiver available "out of the box". It would be hard to find an R8B owner who is unhappy with the radio. If your radio budget permits, the R8B is an excellent receiver and it will not disappoint. Can anything be heard on a stock R8B that cannot be heard on a stock R75? No. Albeit the stock R75 may need to be tuned using ECSS [SSB-mode] instead of SAM-mode for AM [double sideband with carrier] transmissions. Does this matter? Some. SAM normally offers better fidelity than ECSS. A Kiwa Electronics modified R75 with a fixed AM/SAM section will cost a total of $590.

Some may conclude: "Why alter what ICOM should have done right at the factory". While others may conclude: "The R75 can be ordered and shipped directly to Kiwa Electronics: at a savings of $880." In the end, each individual must make that decision. Most R8B owners who have also purchased and modified the R75 are very pleased with their R75's performance. Part of the impetus behind this article was the realization that many are unaware that the R75 modifications even exist. The R75 community identified and corrected flaws. These mods are now professionally offered through Kiwa Electronics. This appears to have went on "under the radar" of the professional reviewers.

Ultimately there is no such thing as a perfect receiver. Both the Drake R8B and ICOM R75 have poorer specifications than, for example, a $1000 used Racal 6790. They both, however, offer more standard features. Ironically, the antenna, often $10 worth of wire, is as important as the receiver itself. The good news is that you will probably be happy with either a Drake R8B or a modified ICOM R75.

3.1 Portable Shortwave Radio Guide
©2011

Shortwave Picks

1. INTRODUCTION

This guide will help you pick a top shortwave radio. Most radios are not worth purchasing. This includes image-laden, single-conversion radios that lack SSB (ex. KA105, FR600, YB-300PE, ATS-404, E10, and CCRadio SWP) and drifty analog-tuned radios with digital frequency-counter displays (ex. KA008, ETFR, GM400, G1100, PT-50, DE1104, S350DL, and S450DLX) and analog dial radios. My picks are show inverted; negatives are gray; and deal-breaking negatives are ~~gray strike~~.

2. DUAL-CONVERSION, PLL-SYNTHESISZED, SIDEBAND-SELECTED AM

The charts only show dual-conversion (for gain distribution and image rejection) and PLL-synthesized (for stability and memories) radios. You can minimize selective fading distortion and select the cleaner AM sideband (for reduced adjacent interference) using **SIDEBAND-SELECTED AM**.

For **SIDEBAND-SELECTED AM** select: AM-mode (slow AGC), a narrow filter (3 or 4 kHz), and offset tune by plus (USB) or minus (LSB) *nearly half* the filter's bandwidth (1 or 2 kHz). You will notice a boost in audio fidelity. Ex. WWCR at 4840 kHz is USB tuned at 4842 kHz and LSB tuned at 4838 kHz. Under *normal* selective fading **SIDEBAND-SELECTED AM** works well; and during heavy fading it is *superior* (there is no lock to be lost) to production portables having Synchronous-AM.

Radio	RF	BW	Step	SSB	Cost	Size	Mem	Knob	Rech
KA11	FAIR	1	~~5 kHz~~	~~NO~~	$50	13	1000	NO	NO
DE1105	FAIR	1	1 kHz	~~NO~~	$60	13	1000	WHEEL	YES
G6 Aviator	FAIR	1	1 kHz	10 Hz	$70	18	700	WHEEL	$8
KA1101	GOOD	2	~~5 kHz~~	~~NO~~	$65	19	20	NO	YES
PL-450	GOOD	2	1 kHz	~~NO~~	$67	21	500	YES	YES
KA1102	GOOD	2	1 kHz	YES	$69	22	133	NO	YES
DE1103	GOOD	2	1 kHz	YES	$72	31	268	YES	YES
DE1121 MP3	FAIR	2	1 kHz	~~POOR~~	$105	34	400	YES	YES
PL-600	FAIR	2	1 kHz	~~POOR~~	$70	40	500	YES	YES
PT-80	FAIR	1	1 kHz	~~POOR~~	$148	52	18	WHEEL	NO
G4000A	GOOD	2	1 kHz	~~POOR~~	$88	60	40	NO	NO
ATS-909X DSP	GOOD	2	1 kHz	40 Hz	$245	62	352	YES	$8
ATS-505P	FAIR	~~1~~	1 kHz	~~POOR~~	$103	65	18	YES	NO
RP2100	GOOD	2	1 kHz	$52	$112	229	30	YES	$8
Satellit 750	GOOD	2	1 kHz	YES	$260	610	600	YES	NO

RF (radio and audio performance), BW (bandwidths), Size (in cubic inches)
Mem (shortwave memories), Rech (recharge system and batteries)
Kaito imports Degen radios; for example, the KA1103 and DE1103 are interchangeable.
*The **CCRadio SW** is the RP2100. The **Tecsun S-2000** is the Satellit 750.*
*Quality control on the **Grundig Satellit 750** and **Grundig G6 Aviator** is not good.*
*The discontinued **ATS-606AP**, **ATS-818ACS**, **E5**, **G5**, and **SW35** are not recommended.*

The **KA1102** is a small and lightweight radio with good audio, decent SSB, and "*automatic tune scanning*" (ATS). Its ATS will automatically populate 20 memories with stations it finds. The radio is good on FM and its buttons are backlit. SSB is only available when in "Page 9" memories.

The **DE1103** is a top modern DXing portable with good whip sensitivity and good immunity to external antenna overload. Its *"pause and continue"* scanning helps to find stations. The knob can be used to step through the memories. The radio does well on FM and its buttons are backlit. Ergonomics are not stellar. The radio has no 5-kHz tuning knob step and the tuning knob doubles as a volume control (press "VOL" then turn the dial; or type a volume number then press "VOL").

The RP2100 is easy-to-use and sensitive. It has a big knob, display, and 5 Watts of audio. The unit has no numeric keypad, no SSB ($52 extra), and only 10 memories per 10 MHz segment.

3. SIDEBAND-SELECTABLE SYNCHRONOUS AM

SAM detectors insert a carrier that maintains phase with the incoming carrier; to help with carrier dropouts. No detector can recover audio lost in a sideband dropout. Distortion is primarily reduced via bandwidth limiting (audio phasing can suppress an entire sideband) and a slow AGC.

Diode detection of AM is forgiving of local oscillator instability: the carrier, akin to being a BFO, moves with the sidebands in perfect "synchronization". During carrier dropouts, signals will appear over-modulated; but, with normal modulation, a diode distorts less than a SAM detector.

Radio	RF	BW	Step	SSB	Cost	Size	Mem	Knob	Rech
G3	FAIR	2	1 kHz	YES	$150	30	700	YES	$8
PL-660	GOOD	2	1 kHz	YES	$110	43	1400	YES	YES
SW7600GR	GOOD	1	1 kHz	YES	$145	45	100	NO	NO

The **DE1106**, possibly similar to the G3, was not available in the US.
Quality control on the **Grundig G3** is not good.
The SW7600GR **holds lock** slightly better than the PL-660 and much better than the G3.

The **SW7600GR** is a *well-built* radio with SAM, good sensitivity, good FM, a slick antenna, and great SSB due to its audio phasing. Sony includes a wind-up antenna (AN-71) and a quality carrying case. The audio is a little *tinny* (hollow) sounding and an AC adapter will run about $20.

4. SILICON LABS Si4734 BASED DSP

The Si4734 is an innovative low-IF DSP receiver chip made by Silicon Labs. I have written five articles on this chip. Unfortunately, some makers are not properly programming the chip. The Si4734 has 5 digital filters (*1, 2, 3, 4, and 6 kHz*), an on-chip s-meter, and it automatically tunes its input. The Si4734 does not decode SSB. Good SSB helped distinguish the radios chosen above.

Radio	RF	BW	SM	SSB	Cost	Size	Mem	Knob	Rech
DE15	FAIR	1	LIKELY	NO	$41	6	100	NO	YES
DE1125[MP3]	FAIR	1	LIKELY	NO	$64	6	100	YES	YES
DE1123[MP3]	FAIR	1	LIKELY	NO	$49	6	100	NO	YES
PL-360	GOOD	1	YES	NO	$58	13	250	WHEEL	YES
PL-606	GOOD	4	YES	NO	$53	13	250	WHEEL	$4
PL-380	GOOD	5	LOW	NO	$53	18	250	WHEEL	YES
PL-300WT	FAIR	1	YES	NO	$55	18	200	WHEEL	YES
G8	FAIR	1	YES	NO	$50	18	200	WHEEL	NO
PL-310	GOOD	5	YES	NO	$53	22	200	YES	YES
PL-390	GOOD	5	LOW	NO	$66	29	250	WHEEL	YES

SM (soft muting): can cause a weak station's audio to drop off or decrease.
Only the **PL-380**, **PL-310**, and **PL-390** have a numeric keypad.

The **PL-390** is a good Si4734-based radio. It has all five digital filters, minimal *soft muting* (can attenuate DX), and a numeric keypad. The ATS/ETM makes finding stations simple. The *line in jack* can be used to play other devices through the dual speakers. The display shows signal-to-noise ratio (dB), signal strength (dBu), and temperature. Tuning step size is *1-Khz*. The **PL-380** is the PL-390's smaller sibling with one speaker, no *line in jack*, and no *antenna jack*. The PL-310 is a popular radio: it is sensitive and has a tuning knob. The PL-310 does have heavier soft muting. Silicon Labs is now the leader in shortwave technology. And Tecsun has a keen eye for ergonomics.

5. MISCONCEPTIONS AND TIPS

You can hear distant stations (DX) using portables, such as those I recommend. A $500 tabletop is not needed; but they do work well, especially on SSB. Using **SIDEBAND-SELECTED AM** you can program listen (SWL) without a SAM detector. In 2003, I proved that an $11 Sony ICF-S10MK2 and Radio Shack Loop could detect **94%** of what an ICOM R75 and Quantum QX Loop could (on MW). DX varies moment to moment and catches often depend upon timing, location, operator skill, luck, and the antenna. I routinely use single-transistor (and triode) radios to DX.

Your first upgrade, before a tabletop, should be the antenna. In urban locations tape 50 feet of wire wrap (nearly invisible) high up on your wall. Another option is to build a tuned loop. Loops reduce mixer energy and can be rotated to stop local noise. *When tuning SSB on a radio without an USB/LSB switch, try offset tuning by 1 to 2 kHz (USB upward, LSB downward).* The stock speakers on portables are not optimal. Earphones or headphones can greatly enhance the radio's audio. If possible, buy radios from US dealers so that a problem can be returned easier.

6. PORTABLE SHORTWAVE RADIO PICKS

I recommend the KA1102, DE1103, SW7600GR, PL-390, and PL-380. Schematics for the first three are available on the internet. The KA1102 is a great little radio for shortwave listening. The DE1103 is the winner of many modern portable DX contests. The SW7600GR is built to last, and has a good synchronous detector and SSB. The PL-380 and PL-390 are DSP radios with five digital filters, an easy-tuning-mode (finds stations for you), a good display, superb FM, and top-notch ergonomics. You can offset tune these two DSP radios by up to an entire filter bandwidth. Tecsun and Silicon Labs are revolutionizing MW, SW, and FM radio: the PL-390 is sure to please.

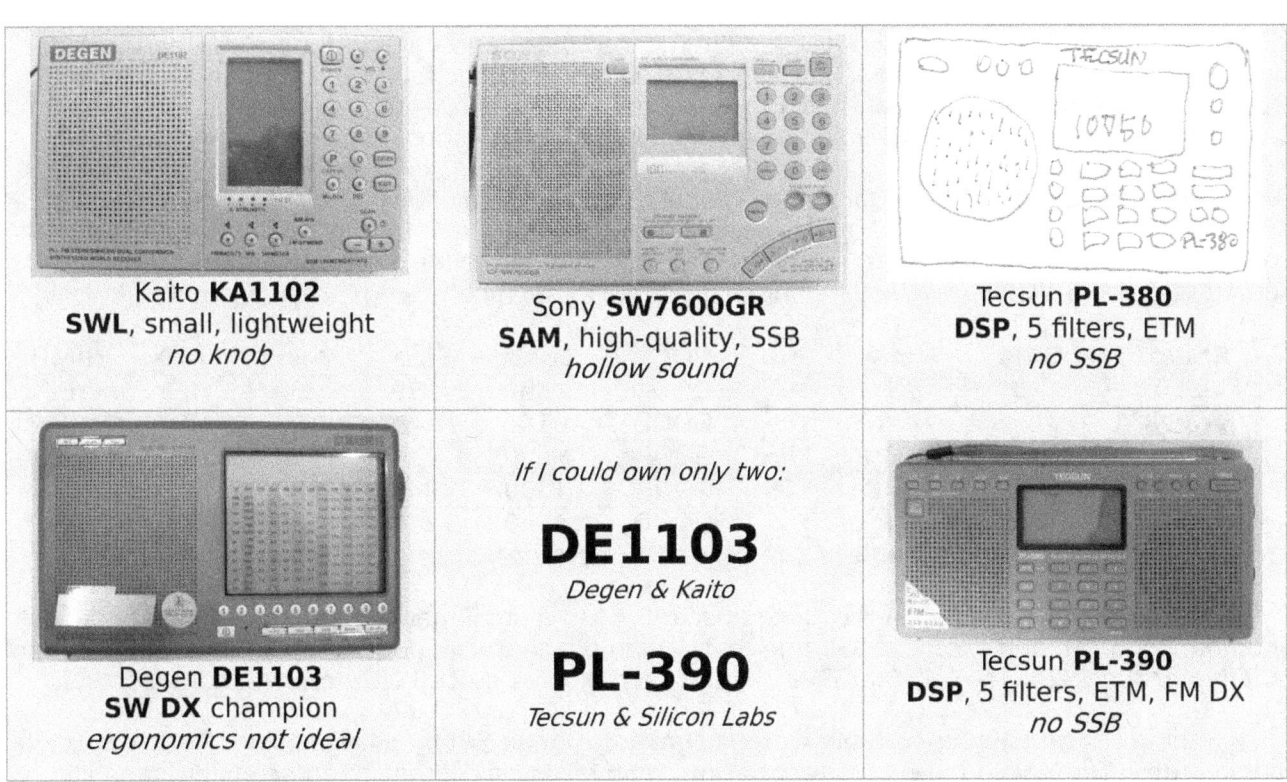

Kaito **KA1102**
SWL, small, lightweight
no knob

Sony **SW7600GR**
SAM, high-quality, SSB
hollow sound

Tecsun **PL-380**
DSP, 5 filters, ETM
no SSB

Degen **DE1103**
SW DX champion
ergonomics not ideal

If I could own only two:

DE1103
Degen & Kaito

PL-390
Tecsun & Silicon Labs

Tecsun **PL-390**
DSP, 5 filters, ETM, FM DX
no SSB

3.2 Phil's SW Radio Picks
©2008

Shortwave Picks

1. DOUBLE-CONVERSION PLL-SYNTHESIZED

The charts *only* include double-conversion (image rejection and gain distribution) and PLL-synthesized (stability and memories) radios. Excluded were all poor performing single-conversion radios lacking SSB (ex. E10, E100, YB-300PE, YB-550PE, YB P2000, KA105, and ATS-404), drifty analog-tuned receivers with digital frequency-counter displays (ex. Mini 300PE, G1000A, G1100B, YB-50, and S350DL), and analog-dial receivers. Charts are organized by **Size** in cubic inches. The "**Mem**" columns reflect only shortwave (SW) memories. Radio picks are show inverted; negative attributes are gray; and deal-breaking negatives (ex. quality control [QC] problems) are ~~striked~~.

2. SAM SIDEBAND-SELECTABLE SYNCHRONOUS AM

SAM detection inserts a carrier that 1) maintains phase with the incoming carrier and 2) in theory is present during selective fades (sideband and carrier dropouts). No detector can recover audio lost to a sideband dropout. Most real SAM detectors sound decent during light carrier fades. Much of the reduced distortion has nothing to do with SAM detection: it entails bandwidth limiting (suppression of one sideband) and using a 2 second release slow-AGC (fast-AGC is for scanning). SAM detectors limit bandwidth via audio phasing (sideband-selection); this results in good fidelity.

Diode detection of AM is forgiving of local oscillator instabilities: the carrier, akin to being a BFO, moves in step with the sidebands (perfect "synchronization"). During carrier dropouts signals appear over-modulated; however, during normal modulation diodes distort less than the best SAM detectors. Good SWL (shortwave listening) is possible in AM-mode using a trick presented below.

Radio	SAM	Audio	SSB	QC	Cost	Size	Mem	Knob	NiMH
KK-S500	FAIR	FAIR	NO	GOOD	$90	29	200	YES	YES
SW7600GR	FAIR	FAIR	YES	GOOD	$146	45	100	NO	N/A
E1	GOOD	GOOD	YES	~~POOR~~	$500	214	1700	YES	N/A
7030+	FAIR	GOOD	YES	~~POOR~~	$1500	348	400	YES	N/A
RX340	FAIR	FAIR	YES	GOOD	$4250	1240	100	YES	N/A

Not being capable of charging NiMH batteries is denoted by "N/A".

The **SW7600GR** is a classic: inexpensive SAM in a small package, good sensitivity, a slick antenna, good FM, and excellent SSB due to its audio phasing. The downside: audio is *tinny* and an AC adapter runs $20. The SW7600GR is an unusual mix of advanced features and hollow sound.

The E1 is an alluring portable with its Drake designed SAM, audio phasing, PBT, and 10 Hz tuning steps. The easy-to-use E1 offers an informative display, great band scanning capabilities, separate tone controls, a whip antenna, FM BCB reception, and can be powered for ~71.4 hours using 4 "D" type alkaline batteries. The downside: no MW ferrite rod, no carry handle, and no 5-kHz knob tuning step. Unfortunately the E1 has emerging QC problems including a safety recall, LCD failures, AC adapter problems, etc. The E1, the top performer, is no longer a pick due to QC.

The KK-S500 has no SSB capability. The 7030+ has prior QC issues and its SAM produces low level *heterodynes* due to harmonic mixing at the "SYNC CAR MIXER". DXer *Dallas Lankford* calls the 7030+ SAM "unacceptable". Reviewer *Dave Zantow* calls the RX-340 SAM "*almost worthless*".

3. PORTABLES: SIDEBAND-SELECTED AM TRICK

SIDEBAND-SELECTED AM minimizes selective fading distortion and permits selection of the cleaner AM sideband (reduces adjacent interference). For **SIDEBAND-SELECTED AM** select: AM-mode, slow-AGC, a narrow filter (~4 kHz), and offset tune (or apply PBT) by plus (USB) or minus (LSB) nearly half the filter's bandwidth (~2 kHz). The audio will brighten as fidelity increases. The BBC at 5975 kHz is USB tuned at 5977 kHz and LSB tuned at 5973 kHz. Under *normal* selective fading **SIDEBAND-SELECTED AM** works well and is doable on portables with a narrow filter and 1-kHz steps. During heavy fading **SIDEBAND-SELECTED AM** is *superior* to SAM because there is no lock to be lost.

Radio	RF	BW	Step	SSB	Cost	Size	Mem	Knob	NiMH
KA1105	FAIR	1	1 kHz	NO	$77	13	1000	YES	YES
KA11	FAIR	1	~~5 kHz~~	NO	$50	13	1000	NO	N/A
KA1101	GOOD	2	~~5 kHz~~	NO	$66	19	20	NO	YES
DE1102	GOOD	2	1 kHz	YES	$64	22	133	NO	YES
ATS-606AP	FAIR	~~1~~	1 kHz	NO	$120	26	54	NO	N/A
E5/G5	GOOD	2	1 kHz	YES	$150	31	700	YES	$18
DE1103	GOOD	2	1 kHz	YES	$64	31	268	YES	YES
KA1121	FAIR	2	1 kHz	POOR	$150	34	400	YES	YES
SW35	FAIR	~~1~~	1 kHz	NO	$90	40	50	NO	N/A
YB-80	FAIR	1	1 kHz	POOR	$160	52	18	YES	N/A
G4000A	GOOD	2	1 kHz	POOR	$150	60	40	NO	N/A
ATS-505AP	FAIR	~~1~~	1 kHz	POOR	$120	65	18	YES	N/A
ATS-909	GOOD	2	1 kHz	40 Hz	$240	70	307	YES	N/A
KA2100	GOOD	2	1 kHz	$33	$100	229	30	YES	$18
ATS-818ACS	FAIR	2	1 kHz	POOR	$200	236	18	YES	N/A

The DE1102 and DE1103 are classics: earphones, AC adapter, rechargeable batteries, long wire antenna, wrist loop, and cloth pouch are included. They both have backlit buttons and worthy FM reception. The **DE1102** is a miniature radio with good SSB, decent audio, and "*automatic tune scanning*" (populates memories). The downside: the DE1102 has SSB only in "Page 9" memories.

The **DE1103** combines superior whip sensitivity with decent immunity to external antenna overload. The DE-1103 has "pause and continue" band scanning, a simulated analog display, and memories can be stepped through via the tuning knob. The downside: the DE1103 has no 5-kHz tuning knob step and the tuning knob doubles as the volume control (press "VOL"; turn the dial).

The **KA2100** is a hefty radio with good MW and SW sensitivity, a big tuning knob, a large display, and powerful audio (2.5 Watts). The downside: it has no keypad, SSB is via an external unit, and only 10 hard-to-recall memories per 10 MHz segment. Knob twisters will like this radio.

The expensive ATS-909 has good SSB but poor whip sensitivity and easily breakable parts (stand and battery cover). The G4000A (was YB-400PE) has decent audio but poor SSB and needs six "AA" type batteries. The E5/G5 improved the DE1103's ergonomics but with a performance hit.

Radios which cannot do **SIDEBAND-SELECTED AM** (5-kHz steps or poor narrow filter) include the KA11, KA1101, ATS-606AP, SW35, and ATS-505AP. The mediocre ATS-818ACS requires 4 "D" and 3 "AA" batteries. The KA1121 has a tiny display and poor ergonomics. The YB-80 suffers from AC adapter hum. The KA1105 is a decent ultra-portable (13 in³) but only covers down to 5.8 MHz.

4. TABLETOPS: PRECISION ECSS TRICK

PRECISION ECSS quickly and accurately tunes the BFO to within 2-Hz of the station's carrier. This is ECSS: tuning AM signals as SSB. Small tuning steps (1-Hz) and high stability (±1 ppm) are ideal: stations are quite stable. ECCS requires no carrier lock and works well on faint signals (DX). For **PRECISION ECSS** select: SSB-mode, a slow AGC, wide filter (~6 kHz) [both sidebands are heard simultaneously], and tune in 1-Hz steps until there is no "flutter" [no discrepancy between the BFO and station carrier]. Afterwards engage a normal SSB filter (~2.5 kHz) for program listening.

Under *heavy* selective fading **PRECISION ECSS** is *superior* to SAM because audio is disturbed less during carrier drops. SSB mode (ECSS) is utilized for reception of low powered signals (called DX: distance reception) due to higher sensitivity and lack of need for lock (SAM is not ideal). DX radios preferably have fine tuning steps, high stability, good filters, and multiple tools. DX varies constantly: reception depends on many factors such as the antenna, time, location, skill, luck, etc.

DDS-SYNTHESIZED MULTIPLE-CONVERSION TABLETOP RECEIVERS						
Feature	RX340	746Pro	7030+	R75.12	E1	R30CC
Steps in Hz	1	1	2.7	1	10	20
±ppm/hr	1	1	1	1	10	5
Filters	57	102	4□	4+□*	3+*	2
Sideband Selection	16-bit ADC DSP	24-bit ADC DSP	FILTER	FILTER	FILTER PHASING	FILTER
Quality Control	GOOD	GOOD	~~POOR~~	GOOD	~~POOR~~	~~POOR~~
Cost	$4250	$1500	$1500	$500	$500	$660
Sensitivity	0.25 µV	0.16 µV	0.19 µV	0.16 µV	0.25 µV	0.50 µV
DR	~~POOR~~	GOOD	GOOD	GOOD	FAIR	GOOD
PBT	YES	DUAL	YES	DUAL	YES	NO
DSP NR	YES	YES	$240	YES	$240	$240
Noise Blanker	YES	YES	$340	YES	NO	NO
Notch	YES	YES	$340	YES	NO	NO
RF Gain	YES	YES	YES	YES	NO	NO
Attenuator	YES	YES	YES	YES	NO	YES
Memories	100	100	400	99	1700	100
Display	FAIR	GOOD	FAIR	GOOD	GOOD	FAIR
Tuning Knob	GOOD	GOOD	GOOD	GOOD	LIGHT	LIGHT
Keypad	YES	YES	REMOTE	YES	YES	NO
Speaker	POOR	POOR	GOOD	POOR	GOOD	POOR
RS232	YES	YES	YES	YES	NO	NO
Size (inches³)	1240	665	348	317	214	163

□*Accepts an additional 455-kHz IF filter: ex. Collins mechanical with 100 dB ultimate rejection.*
**The R75 twin-PBT and E1 phasing-PBT allow the analog simulation of multiple filter bandwidths.*

The **R75** stands out for DX due to its 1 Hz tuning steps, ±1 ppm stability, dual-PBT, high sensitivity, numerous DX tools, and build quality. *WRTH* named the R75 "Best Value Tabletop" back when it cost $1040; and *Passport* calls the R75 *"first-rate for unearthing tough utility and ham signals"*. The R75, achieving many bandwidths via the twin-PBT, needs no additional filters.

The RX340 has poorer blocking, ultimate rejection, and dynamic range than the R75. DXer *Jan Alvestad* found his modified R75 to have better sensitivity and audio quality than the RX340.

The 7030+ has no bandpass filters (increased mixer trash), no DDS shielding, some SSB *hiss*, and prior QC problems. The noise blanker and notch cost an additional $340. *Jan Alvestad* found his modified R75 to have superior sensitivity and a better frontend (BPFs) than the 7030.

The E1 is a capable portable but its build quality (mechanical encoder, buttons, overall) is not equal to tabletop standards. Display contrast, IP3 values, and DX tools are better on the R75. The E1 has no DSP noise reduction, no CPU control, and has been plagued with rising QC issues.

The R30CC offers good audio, an analog s-meter, and is ideal for portable DX using large antennas. However, its 10 "AA" batteries are hard to replace and last only ~8.1 hours. The R30CC is missing squelch, scan, and clock functions. The R30CC has no CPU control and some QC issues.

Guy Atkins compared his modified R75 with his custom professional Racal RA6790GM. He found the R75 *"more flexible in tough DX situations than the RA6790GM"*. He noted that the R75 was easier to operate, better on severely overlapping frequencies, and able to peak crucial voice frequencies via the twin-PBT for best intelligibility. Guy stated that: *"the stock filtering and twin-PBT are a powerful combination"*. The RA6790GM receiver sold new for $6000 in the 1980's.

5. DISCUSSION

I highly recommend the DE1102, DE1103, KA2100, and R75 (±SE-3). Six tabletops have been discontinued lately: the R8B, RX-350, Sat800, FRG100, NRD545, and R75-02 (SAM version). Upcoming portables include the S-2000, KK-S7600L, and RP3100; the later two have SAM-mode.

SAM detectors in the RX-340, 7030+, SW7600GR, and KK-S500 regularly lose lock during *heavy* carrier fades. The E1 maintains lock but sounds strained. ECSS on the E1 is top-notch due to its mix of phasing and 10-Hz steps. Regrettably, the E1 is no longer a pick: like its Satellit 800 brethren, it has mediocre build quality and mounting QC issues. SW7600GR audio is *tinny* but its SAM is similar to more costly units which can whistle, hiss, and strain to gain and maintain lock. These SAM units do not live up to one inside the R8B. Tuning tricks make SAM more superfluous.

AM-mode can reduce selective fading distortion by using a slow-AGC with an offset-tuned narrow filter (**SIDEBAND-SELECTED AM**). Consumers are unaware that most SAM detectors function poorer than this easily done tuning trick. Ideal portables are: the DE1102 for SWL and DE1103 for DX. The KA2100 is good for MW DX. Makers should add a U/L labeled button that shifts the LO without varying the display frequency: slow-AGC and narrow filter would automatically be chosen.

DSP-IF receivers offer numerous digital bandwidths; nonetheless, their 8-bit "mechanical" sound is not ideal for SWL. The **746Pro** has a *higher* dynamic range (70 dB @ 2 kHz) and contains *better* [24-bit ADC] DSP-IF receive section than the RX340, NRD545, or WJ-8711A. If DSP-IF is a must, get the 746Pro. Look for potential ham deals: the unit is known for transmit section failures.

The R75 is an idyllic tabletop with its 1-Hz tuning and ±1 ppm stability (**PRECISION ECSS**). If SAM is a must, the R75 can be mated to *Robert Sherwood*'s (*www.sherweng.com*) $570 **SE-3** MK III D High-Fidelity Phase-Locked AM Product Detector via the $35 Dual IF SE-3 Mod and $45 BUF-3 Output Amp [$45 for installation]. The $1195 combination of SE-3 and R75 is economical compared to a similarly equipped 7030+ (notch, NB, and DSP NR) at ~$2080. The SE-3 SAM is perhaps the best SAM ever implemented. The R75 mated with SE-3 SAM remains a *killer combo*.

Antennas are vital for good reception. Homebrew amplified tuned-loops work well in urban locations. They lower NF, reduce mixer trash, and can be rotated to attenuate local noise sources.

Radio Pick	Cost	Distinction
Sony SW7600GR	$146	VALUE SAM
Degen DE1102	$64	VALUE SWL
Degen DE1103	$64	VALUE DX
Redsun KA2100	$100	LARGE MW DX
ICOM IC-R75-12	$500	TABLETOP DX (±SE-3 SAM)
ICOM IC-746Pro	$1500	TABLETOP DSP-IF

3.3 Phil's SW Radio Buying Guide
©2006

Portable Review

 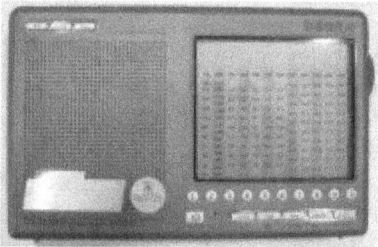

1. INTRODUCTION

This guide *only* includes well performing portables. Among the poor performers were numerous single conversion units lacking SSB and drifty analog units with digital frequency counter displays (no memories). Note: the S350DL is decent on MW (large ferrite rod). Portables are organized by **Size** in cubic inches. The "**Mem**" columns below reflect only SW memories.

2. DOUBLE CONVERSION

Double conversion helps reduce images. Portables without audio phasing (ss-SAM) need two bandwidths: to reduce adjacent interference simply engage the narrow filter and detune by one or two kHz. Without SAM, good SSB is necessary to reduce heavy selective fading distortion.

Radio	Rating	Cost	Size	Mem	SSB	BW	Knob	AC-Bat
KA1101	GOOD	$66	23	20	NO	2	NO	YES-NiMH
DE1102	GOOD	$64	24	133	YES	2	NO	YES-NiMH
E5	GOOD	$150	31	700	YES	2	YES	YES-NO
DE1103	GOOD	$64	32	268	YES	2	YES	YES-NiMH
YB-80	GOOD	$160	52	18	POOR	1	YES	YES-NO
YB-400PE	GOOD	$130	60	40	POOR	2	NO	YES-NO
G4000A	GOOD	$150	60	40	POOR	2	NO	YES-NO
ATS-909	GOOD	$240	69	307	40 Hz	2	YES	YES-NO

Positive attributes are green; negative are yellow; and deal-breakers are red.

The KA1101 has few memories and no SSB. The E5, YB-80, and G4000A cost more than the SW7600GR which has SAM. The YB-400PE has good audio but poor SSB and needs six "AA" type batteries. The costly ATS-909 has good SSB but poor whip sensitivity and easily breakable parts (stand and battery cover). Some E5 units are not as sensitive as the DE1103. Missing from the chart are the mediocre performing SW35, ATS-505P, ATS-606AP, and ATS-818ACS.

The DE1102 and DE1103 are classics: earphones, AC adapter, rechargeable batteries, external antenna, and cloth pouch are included. The DE1102 is an ultra-portable with good SSB (for reduced selective fading distortion). The DE1103 combines good whip sensitivity with good immunity to external antenna overload. The downside: the DE1102 has SSB only in the "Page 9" memories. The DE1103 has no 5-kHz tuning knob step and the tuning knob doubles as the volume control. I recommend the purchase of the Degen radios through quality sellers on eBay.

Distinction	Portable	Cost
GOOD VALUE SWL	Degen DE1102	$64
GOOD VALUE DX	Degen DE1103	$64

3. DOUBLE CONVERSION WITH SAM

SAM (Synchronous AM detection) enhances audio by inserting a carrier that maintains phase with the incoming carrier. SAM also reduces the distortion caused by selective fading.

Radio	Rating	Cost	Size	Mem	SSB	BW	Knob	AC-Bat
SW7600GR	GOOD	$146	45	100	YES	1	NO	NO-NO
E1	GREAT	$500	244	1700	10 Hz	3+*	YES	YES-NO
Sat800	GREAT	$500	1641	70	50 Hz	3	YES	YES-NO

The E1 phasing-PBT allows the analog simulation of multiple filters.

The discontinued Sat800 is a great performer but too large and heavy (15 lbs.) for easy portability. Reconditioned units are available for $420.

The SW7600GR is a classic: inexpensive SAM in a small package and great SSB due to its audio phasing. The downside: audio is tinny; and rechargeable batteries and AC adapter are not included. A generic charge unit with batteries will cost about $18 while AC adapters run about $20.

The E1 is the highest performance portable available with: audio phasing, PBT, and 10 Hz tuning steps. The E1 has a Drake engineered SAM, informative display, and great band scanning capabilities. The downside: no MW ferrite rod, no carry handle, and no 5-kHz knob tuning step.

Distinction	Portable	Cost
INEXPENSIVE SAM	Sony SW7600GR	$140
TOP PERFORMANCE	Eton E1	$500

4. DISCUSSION

There are numerous portables. The four I recommend will suit most people: the DE1102, DE1103, SW7600GR, and E1. The DE1102 is good for casual SWL and the DE1103 is good for casual DX. Both come with rechargeable batteries for convenient portable usage. The SW7600GR features good sensitivity, excellent SSB, and SAM mode for pleasant broadcast listening. The E1 portable has near tabletop level performance. E1 units numbered 3067 to 5462 were recalled: there is a possibility of battery rupture and leakage.

Distinction	Portable	Cost
GOOD VALUE SWL	Degen DE1102	$64
GOOD VALUE DX	Degen DE1103	$64
INEXPENSIVE SAM	Sony SW7600GR	$140
TOP PERFORMANCE	Eton E1	$500

Tabletop Review

1. INTRODUCTION

Feature	RX340	NRD545	7030+	RX350	FRG100	R30C	R75	E1
Cost	$4250	$1800	$1500	$1200	$600	$590	$570	$500
Size*	1247	944	348	788	301	163	320	244

*Size is in cubic inches. Negative attributes are gray and deal-breakers are ~~gray striked~~.

Recently Drake discontinued their venerable R8B and Eton released their E1. The E1, a high performance portable, replaces the quality control (QC) problem laden and discontinued Sat800 in the charts below. Tabletops are organized by **Cost**. Each radio has its own niche.

2. SWL: SHORTWAVE LISTENING

SWL involves reception of relatively high powered broadcast signals. SAM mode enhances reception by inserting a carrier that maintains phase with the incoming carrier. SAM reduces the distortion caused by selective fading. All the radios below can be used for SWL.

Feature	RX340	NRD545	7030+	RX350	FRG100	R30C	R75	E1
SAM	FAIR	FAIR	GOOD	~~POOR~~	~~NONE~~	~~NONE~~	GOOD	GOOD
Audio	GOOD	~~POOR~~	GOOD	GOOD	GOOD	GOOD	GOOD	GOOD
Speaker	POOR	POOR	GOOD	POOR	POOR	POOR	POOR	GOOD
Memory	100	1000	400	1024	50	100	99	1700
Display	GOOD	GOOD	FAIR	GOOD	FAIR	FAIR	GOOD	GOOD

The R75 requires modification: the stock SAM and audio are POOR.

The E1 stands out for SWL due to its Drake designed SAM, pleasant audio, big speaker, numerous memories, and informative display. The easy-to-use E1 has separate tone controls, a whip antenna, FM band reception, and can be readily battery powered.

The $1500 7030+ is excellent for SWL but with prior QC issues. At one third the price the E1 has a superior display, better ergonomics, and more memories.

The R30C can be battery operated but batteries are hard to replace and last only 8.1 hours. The E1 can run for 71.4 hours on alkaline batteries. The R30C has some QC issues.

The R75 has an overly fast SAM AGC, limited fidelity, and a broken SAM. *Kiwa Electronics* can fix these problems for $80 via two modifications: 1) Synchronous Detector Upgrade, and 2) High-Fidelity Audio Filter Upgrade. The R75 SAM is sideband selectable via the 9-MHz IF filter. The E1 and 7030+ need no modification for great SWL.

Usage	Tabletop	Cost
SWL	Eton E1	$500
SWL	AOR 7030+	$1500

3. DX: DISTANCE RECEPTION

DX involves reception of low powered signals such as hams, utilities, or broadcasts. SSB mode (ECSS: tuning DSB signals as SSB) is used due to higher sensitivity and lack of need for lock. DX receivers ideally have fine tuning steps, high stability, good filters, and multiple tools. DX varies constantly: reception depends on factors such as the antenna, time, location, skip, gray line, sunspots, solar cycles, skill, luck, etc. All the radios below can be used for DX.

Feature	RX340	NRD545	7030+	RX350	FRG100	R30C	R75	E1
Steps Hz	1	1	2.7	1	10	20	1	10
±ppm/hr	1	2	1	1	10	5	1	10
Filters	57	998	4*	34	3*	2	4+*□	3+□
Sensitivity μV	0.25	0.32	0.19	0.35	0.25	0.50	0.16	0.25
DR	~~POOR~~	GOOD	GOOD	~~POOR~~	GOOD	GOOD	GOOD	FAIR
DSP NR	YES	YES	NO	YES	NO	NO	YES	NO
Noise Blanker	YES	YES	$340	FAIR	YES	NO	YES	NO
Notch	YES	AUTO		AUTO	NO	NO	AUTO	NO
RF Gain	YES	YES	YES	FAIR	NO	NO	YES	NO
Attenuator	YES	YES	YES	YES	YES	YES	YES	NO
Knob	GOOD	GOOD	GOOD	GOOD	LIGHT	LIGHT	GOOD	LIGHT
Keypad	YES	YES	REMOTE	$140	NO	NO	YES	YES
RS232	YES	YES	YES	YES	$90	NO	YES	NO

□The R75 twin-PBT and E1 phasing-PBT allow the analog simulation of multiple filters.
*Accepts an additional 455 kHz IF filter: ex. Collins mechanical with 100 dB ultimate rejection.

The R75 stands out for DX due to its 1 Hz tuning steps, ±1 ppm stability, dual-PBT, high sensitivity, numerous DX tools, and build quality. *WRTH* named the R75 "Best Value Tabletop" back when it cost $1040; and *Passport* calls the R75 "*first-rate for unearthing tough utility and ham signals*". The R75 comes with DSP noise reduction: aftermarket units can cost $240.

The $1800 NRD545 is plagued with poor audio: hiss, unnatural sound, and DSP noise.

The $4250 RX340 has poorer blocking, ultimate rejection, and dynamic range than the R75. *Jan Alvestad* found his modified R75 to have superior sensitivity and better audio quality than the RX340.

The $1500 7030+ has no bandpass filters, no DDS shielding, some SSB hiss, and prior QC issues. The noise blanker and notch cost an additional $340. *Jan Alvestad* found his modified R75 to have superior sensitivity and a better frontend than the 7030.

Guy Atkins compared his modified R75 with his custom professional Racal RA6790GM. He found the R75 "*more flexible in tough DX situations than the RA6790GM*". He noted that the R75 was easier to operate, better on severely overlapping frequencies, and able to peak crucial voice frequencies via the twin-PBT for best intelligibility. He stated that: "*the stock filtering and twin-PBT are a powerful combination*".

Usage	Tabletop	Cost
DX	ICOM R75	$570

4. DISCUSSION

I previously recommended the now discontinued $1470 R8B: labeling it both the Best Performer for SWL and the Easiest for DX. The RX350D was recently discontinued as well. The R30C remains a good choice for portable DX using large antennas; its audio is very good.

DSP-IF receivers offer many digital bandwidths. Unfortunately, the RX340 and RX350 have inadequate 16-bit ADC lines which result in poor dynamic range. SAM modes on these radios and the NRD545 are mediocre and the NRD545 has poor audio. The 24-bit ADC $1500 ICOM 746Pro has 102 filters, great dynamic range, high sensitivity, and a SAM mode that *Dallas Lankford* calls "*outstanding*" because it never loses lock. This said, I doubt many will buy the 746Pro for SW even though it contains a *better* DSP-IF receive section than the RX340, NRD545, and RX350.

I also doubt many will spend $675 extra to mate the R75 to *Robert Sherwood*'s excellent SE-3 MK III D High-Fidelity Phase-Locked AM Product Detector.

I recommend the E1 or 7030+ for SWL and the R75 for DX. The E1 (audio phasing mated with PBT) and R75 (twin-PBT) each employ analog means to achieve multiple filter bandwidths.

For both SWL and DX I recommend the 7030+, modified R75, or both the E1 and R75. Tabletops costing $1200 to $4250 would be hard pressed to beat the $1070 **E1-R75 combination**. A similarly equipped 7030+ (notch, noise blanker, DSP noise reduction) would cost nearly double: ~$2080. The R75 audio can be run through the E1 auxiliary input to make use of the E1 speaker, volume, and tone controls. For the price-conscious the modified R75 is a potent all-purpose setup. The modified R75 is superior to the E1 in build quality (tuning knob, buttons, overall), display contrast, DSP noise reduction, IP3 values, and DX tools. Those who do mostly SWL and run indoor antennas will likely be happy with the E1. The 7030+ is excellent but expensive. For SWL most of what is out there that is worth listening to hour after hour can be heard just as well using an E1 or modified R75. For DX it is hard to beat the stock R75.

The radios I did not recommend (RX340 and RX350 for dynamic range; NRD545 for audio; FRG100 and R30C for no SAM and few DX tools) are all great radios. However, most would be well served with an E1, 7030+, R75, modified R75, or E1-R75 combination.

Usage	Tabletop	Cost
SWL	Eton E1	$500
SWL	AOR 7030+	$1500
DX	ICOM R75	$570
SWL + DX	Modified ICOM R75	$650
SWL + DX	Eton E1 and ICOM R75	$1070
SWL + DX	Upgraded AOR 7030+	$1840

3.4 Phil's Tabletop Guide 2005

2005

1. INTRODUCTION

This guide reviews the tabletop receivers and exposes their flaws. Each has a niche and overall performance is similar. The chart below is a summary. Significant negative atributes are shown in gray and positives are show in **bold** or are `inverted`.

Feature	NRD545	7030	R8B	RX350
Maker	JRC	AOR	Drake	TenTec
Price USD	$1,800.00	$1,490.00	$1,470.00	$1,200.00
Memories	1000	400	1000	1024
Keypad	YES	REMOTE	FLIMSEY	$140.00
Knob	GOOD	GOOD	LIGHT	GOOD
S-Meter	ANALOG	DIGITAL	ANALOG	DIGITAL
Tone	YES	YES	YES	PBT
FM-Mode	YES	YES	YES	YES
Quality	GOOD	FAIR	GOOD	GOOD
Batteries	NO	NO	NO	NO
Speaker	POOR	**GOOD**	POOR	POOR
Conversion	triple	double	double	triple
Extra MHz	-	-	-	-
Weight	16.5	4.9	13	12
Volume	944	348	901	788

Note: the 7030+ is priced better in Europe. Weight is in pounds; volume is in cubic inches.

Feature	Mod R75	FRG100	R30	Sat800
Maker	ICOM	Yaesu	Palstar	Grundig
Price USD	$580.00	$600.00	$500.00	$500.00
Memories	100	52	100	70
Keypad	YES	NO	NO	FLIMSEY
Knob	GOOD	LIGHT	LIGHT	GOOD
S-Meter	DIGITAL	ANALOG	ANALOG	ANALOG
Tone	PBT	PBT	DETUNE	YES
FM-Mode	YES	$45.00	NO	YES
Quality	GOOD	GOOD	FAIR	FAIR
Batteries	NO	NO	**10 "AA"**	**6 "D"**
Speaker	POOR	POOR	POOR	**GOOD**
Conversion	triple	double	double	double
Extra MHz	30-60 VHF	-	-	118-137AIR 88-108FM
Weight	6.6	6.6	2.2	14.6
Volume	320	301	163	1641

Each radio has an attenuator, variable AGC, and narrow (~2.35 kHz) filter. The R30 alone is missing squelch, scan, and a clock. Four cost significantly less: the R75, FRG100, R30, and Sat800. Three lack a direct frequency entry keypad: the RX350 [$140 extra], FRG100, and R30. Three have lightweight knobs: the R8B, FRG100, and R30. The 7030+ and R30 have undergone silent revisions. 7030+ failures include: buttons, speakers, jacks, encoders, power supply, J309 JFET, SD5400 mixer, etc. Each 7030+ has its individual quirks. R30 failures include: power switch, adapters, encoders, etc. The Sat800 has quality control issues and mediocre construction but is under warranty for 1 year by R.L. Drake. The Sat800 is huge, the R30 is tiny; both can be battery operated. The 7030+ is sold in few stores in North America and the FRG100 has been discontinued.

2. USAGE

Tabletops are used for two distinct reasons. SWL ("shortwave listening") is hour after hour listening to broadcast shows transmitted at high power (aids reception by portables). DX ("distance reception") is identifying low power or distant signals including amateurs and utilities (military, aviation, maritime, beacon).

Signal	Type	Mode	Power	Usage	Publication
SWL	DSB	AM/SAM	High	major programs or news broadcasts	*Passport*
DX	SSB	SSB	Low	utilities, ham, pirate, clandestine	*WRTH*

Note: DSB and SSB are both amplitude-modulated emissions.

3. SWL: SHORTWAVE LISTENING

All of the tabletops are capable of hearing SWL. However, for optimal listening, the receiver needs a good AGC, high fidelity, and synchronous detection (SAM). A slow AGC is critical for dealing with normal fading. SAM clears reception by inserting a carrier that maintains phase with the incoming carrier. SAM also reduces distortion caused by a type of fading called selective fading. ECSS (tuning DSB signals as SSB) can be used for severe selective fading. Under ECSS the BFO does not require a carrier for lock but receiver stability and fine tuning steps are critical.

Since DSB offers two copies of the audio the sideband with the least interference can be selected via phasing or filter. The radios offering SAM can select a sideband; however, only the R8B and Sat800 include R. L. Drake's high-performance and easy to use design. The R75 can select a sideband in SAM mode by engaging the narrow 9-MHz IF SSB filter and applying PBT.

The stock $500 R75 has a too-fast AM/SAM AGC, poor fidelity, and a broken SAM. Design flaws and their fixes are well documented at the *R75 Yahoo Group*. ICOM made electrically minor errors that severely degraded SWL performance. The R75 uses the Motorola C-QUAM® stereo decoder chip [0.3% THD and ± 3 kHz lock range]. Kiwa Electronics (*www.kiwa.com*) can modify an R75 for $80 as follows for optimal SWL:

Price	Kiwa R75 Mod	Fixes
$45	Synchronous Detector Upgrade	SAM and AGC
$35	High-Fidelity Audio Filter Upgrade	Fidelity

The stock $500 R75 can also be mated to *Robert Sherwood*'s (*www.sherweng.com*) $550 SE-3 MK III D High-Fidelity Phase-Locked AM Product Detector via the $35 Dual IF SE-3 Mod and $45 BUF-3 Output Amp [plus $45 for installation]. This potent yet *affordable* combination of high performance DX rig and high performance SAM has received some glowing reports.

SWL and DX	Radio	Cost
KILLER COMBO	Sherwood SE-3 / ICOM R75	$1175

The NRD545 and RX350 are slow to gain and have difficulty maintaining SAM lock. The NRD545 has poor audio: hiss, unnatural sound, DSP burps/clicks, and low volume. The RX350 has firmware bugs which can cause the unit to freeze; resets, unfortunately, erase the memories. The FRG100 and R30 do not have SAM.

Feature	NRD545	7030	R8B	RX350
Price	$1,800.00	$1,490.00	$1,470.00	$1,200.00
SAM	FAIR	AOR	**Drake**	BROKEN
Audio	POOR	GOOD	GOOD	GOOD

Feature	Mod R75	FRG100	R30	Sat800
Price	$580.00	$600.00	$580.00	$500.00
SAM	Motorola (Kiwa)	NO	NO	**Drake**
Audio	GOOD (Kiwa)	GOOD	GOOD	GOOD

Note: the R75 requires modification for optimal SWL usage.

Below is a list of published SAM mode "overall distortion" values.

Feature	RX340	7030+	R8B	Sat800
Price	$3950	$1490	$1470	$500
SAM Distortion	2.6%	2.0%	0.6%	2.4%

4. SWL RECOMMENDATIONS

For optimal SWL the Drake R8B is the clear winner. *Passport* states that the R8B "gets everything right". However most would be happy with a modified R75 or Sat800 at one third the price. Note that *Passport* rated the modified R75 better than the Sat800 without testing the SWL enhancing High-Fidelity Audio Filter modification. The SE-3 enhanced R75 is also an option. *Other tabletops can be mated to the Sherwood SE-3 as well.* The 7030+ offers Sat800 level SAM distortion at triple the cost. The other radios have poor or missing SAM. Ironically the Sat800 has less SAM distortion than the professional $3950 RX340. The Sat800 is ideal for beginners and non-technical buyers as it is easy to use and comes with a built-in antenna and good speaker. The unit should however be examined for defects. Note that the Eton E1XM will likely replace the Sat800.

SWL Distinction	Radio	Cost
BEST PERFORMER	Drake R8B	$1470
BEST VALUE	Modified ICOM R75	$580
KILLER COMBO	Sherwood SE-3 / ICOM R75	$1175
EASIEST	Grundig Sat800	$500

5. DX: DISTANCE RECEPTION

All of the tabletops are capable of hearing DX. However, for optimal monitoring radios need great SSB: fine tuning steps, good filters, and multiple tools. On weak stations SAM is difficult to use: even the Drake and AOR designs emits hiss and have trouble gaining lock. The RX350 alone has a small bandscope; unfortunately, the entire display is pixilated and ghosts. The R75 alone has dual pre-amps and twin-PBT.

Specification	NRD545	7030	R8B	RX350
Price	$1,800.00	$1,490.00	$1,470.00	$1,200.00
Tuning Step (Hz)	1	2.7	10	1
Stability (ppm)	2	1	5	1
Filters	998	4	5++	34
Optional Filters	NO	YES	NO	NO
Sensitivity (µV)	0.32	0.19	0.25	0.35

Specification	R75	FRG100	R30	Sat800
Price	$500.00	$600.00	$580.00	$500.00
Tuning Step (Hz)	1	10	20	50
Stability (ppm)	1	10, 2 ($95)	5	10
Filters	4++	3	2	3++
Optional Filters	YES	YES	NO	NO
Sensitivity (µV)	0.16	0.25	0.5	0.5

Note: the R75's twin-PBT allows simulation of multiple filters. The R8B and Sat800 have phasing.

Fine tuning steps of 1-Hz and stability of 1-ppm are ideal for DX. Fine tuning can be achieved via DDS (Direct Digital Synthesis) or PLL (multi-loop or fractional N synthesizer). JND (Just Noticeable Difference) for the human ear is 1.5 Hz at 500 Hz meaning: a human can detect a difference between a 500.0 Hz and 501.5 Hz tone. JND varies with frequency: JND is 2.9 Hz at 1000 Hz, 5.8 Hz at 2000 Hz, and 8.7 Hz at 3000 Hz. An ECSS (tuning DSB signals as SSB) trick involves using a wide (~6 kHz) filter on SSB and tuning in 1-Hz steps until there is no "flutter" [dead on]. After which the normal SSB filter (~2.35 kHz) can be engaged for monitoring. The human ear acts as a mechanical spectrum analyzer: it responds to the amplitude of the harmonic components but not their phase.

The NRD545 and RX350 are capable of numerous bandwidths via DSP IF. Unfortunately both units have inadequate ADC lines [18-bit NRD545; 16-bit RX350] on their DSP units. Ultimate rejection and dynamic range suffer. The R75 achieves numerous bandwidths via twin-PBT. Any SSB filter from ~400 Hz to ~2400 Hz can be simulated without the drawbacks of DSP IF using low resolution ADC inputs. Placing the filters over one sideband of a DSB signal allows simulation of ~800 Hz to ~4800 Hz. The R75 does not need any additional optional ICOM filters. The R8B comes with a potent filter setup: five LC filters and phasing. Internally the R8B has a single 50 kHz filter network that is manipulated by deQing, like the old "4" line.

Feature	NRD545	7030	R8B	RX350
PBT	YES	YES	YES	YES
RF Gain	YES	YES	YES	FAIR
Noise Blanker	YES	$330.00	GREAT	FAIR
Notch Filter	AUTO		YES	AUTO
DSP Noise Reduction	YES	NO	NO	YES
RS232 Control	YES	YES	YES	YES

Feature	R75	FRG100	R30	Sat800
PBT	DUAL	NO	NO	NO
RF Gain	YES	NO	NO	NO
Noise Blanker	YES	YES	NO	NO
Notch Filter	AUTO	NO	NO	NO
DSP Noise Reduction	YES	NO	NO	NO
RS232 Control	YES	$90.00	NO	NO

The FRG100, R30, and Sat800 are missing basic DX features including PBT, RF-gain, and notch filter. Notch filters eliminate single-pitched noise (heterodynes). Noise blankers, missing on the R30 and Sat800, block impulse noise. DSP noise reduction can make searching for signals more enjoyable and protect hearing. The NRD545 and RX350 use IF DSP while the R75 uses an AF DSP [$140 value, free with coupon]. Aftermarket DSP noise reduction units cost $230 and up.

6. DX RECOMMENDATIONS

For DX the ICOM R75 is an excellent choice. Note that *WRTH* awarded the R75 "Best Value Tabletop" back when it cost $1040; and *Passport* calls the R75 "first-rate for unearthing tough utility and ham signals". The DSP IF receivers above are unimpressive: they all have flawed SAM units and inadequate ADC resolution leading to poor dynamic range. Whereas the 24-bit ADC $1550 ICOM 746Pro has excellent dynamic range and a SAM unit that never loses lock. The 746Pro transceiver is a *better* radio than the RX340, NRD545, and RX350. DXer and radio aficionado *Dallas Lankford* called the 746Pro's SAM "outstanding".

Feature	RX340	NRD545	746Pro	RX350
Maker	TenTec	JRC	ICOM	TenTec
Price (keypad)	$3950	$1800	$1550	$1340
SAM	FAIR	FAIR	GOOD	BROKEN
ADC	16-bit	18-bit	24-bit	16-bit
Dynamic Range	46 dB @ 5 kHz	68 dB @ 5 kHz	70 dB @ 2 kHz	POOR
Filters	57	998	51 x 2	34
Sensitivity (µV)	0.25	0.32	0.16	0.35

DX Distinction	Radio	Cost
BEST DSP IF RECEIVER	ICOM 746Pro	$1550

The 7030+ has a high IP mixer but it alone is also missing the bandpass filters which keep trash out of the first mixer. AOR also stopped using DDS shielding due to PCB stress causing carrier oscillator failures: noise has increased. The 7030+ has minor SSB hiss. The wide shape factor LC filters which give the R8B its mellow SWL audio do little for its DX capabilities. The R75, 7030+, and FRG100 can accept Collins 455 kHz IF mechanical filters with over 100dB of ultimate rejection. *WRTH* gave the R75 higher marks than the R8B for mechanical design, construction quality, and ergonomics. The FRG100, R30, and Sat800 are missing many important DX features. Ironically the R75 has better blocking, ultimate rejection, and dynamic range than the professional $3950 RX340. The R30 is ideal for portable LW/MW/SW DX; however, it consumes batteries and five screws must be removed to replace them.

DX Distinction	Radio	Cost
BEST PORTABLE	Palstar R30	$500

DXer *Jan Alvestad* compared his custom R75 with the 7030 and RX340. He found the R75 had superior sensitivity, a better frontend than the 7030, and better audio quality than the RX340. DXer *Guy Atkins* compared his custom R75 with his custom Racal RA6790GM (a $6000 rig from the 1980's). He found the R75 "more flexible in tough DX situations than the RA6790GM". He noted the R75 was easier to operate, better on severely overlapping frequencies, and able to peak crucial voice frequencies via the twin-PBT for best intelligibility. As far as optional filters he noted: "the stock filtering and twin-PBT is a powerful combination; it's only in the toughest of DX conditions that the replacement filters [INRAD] really show their worth."

No radio clearly beats all the rest as far as DX. Propagation varies from minute to minute. Reception often depends not on the radio but on operator skill, luck, skip, geographic location, time-of-day frequency patterns, gray-line reception, sunspot activity, solar cycles, etc. The antenna is more critical to performance than which tabletop. Experienced DXers know that the tabletops are very similar. These receivers all share a similar topology: multiple-conversion super-heterodyne.

DX Distinction	Radio	Cost
BEST PERFORMER	NONE	-
BEST VALUE	ICOM R75	$500
EASIEST	Drake R8B	$1470

7. SWL and DX RECOMMENDATIONS

For both SWL and DX consider the ICOM 746Pro, modified ICOM R75, Sherwood SE-3 enhanced ICOM R75, or Drake R8B. Anyone considering an RX340, NRD545, or RX350 should look at the 746Pro. Anyone considering a 7030+ should look at the R8B. Anyone considering a Sat800, FRG100, or non-portable R30 usage should look at the R75.

SWL and DX	Radio	Cost
BEST DSP IF RECEIVER	ICOM 746Pro	$1550
BEST VALUE	Modified ICOM R75	$580
KILLER COMBO	Sherwood SE-3 / ICOM R75	$1175
BEST SWL AND EASIEST DX	Drake R8B	$1470

8. SKETCHES AND PICTURES

Below are drawings or pictures of each unit with mention of some individual quirks.

	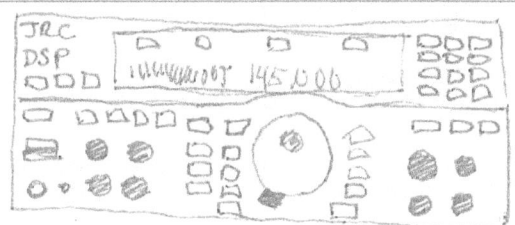
RX350 DUAL VOLUME AND RF-GAIN KNOB	**NRD545** ONLY 10-Hz DISPLAY RESOLUTION
	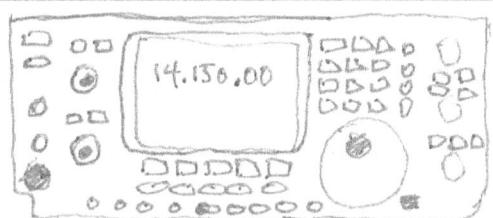
RX340 POOR DYNAMIC RANGE	**746Pro** TRANSMIT SECTION FAILURES

7030+ HEX SCREWS STIP EASILY TREE MENU SYSTEM LARGE BIRDIES	**R8B** VOLATILE MEMORIES, BASSY AUDIO, FUSSY PBT DIAL ENCODER FAILURES, POWER SUPPLY HEAT CW OFFSETS AND WIDE CW FILTER SHAPE

FRG100 CHUFFS WHEN TUNED RAPIDLY	**R30** MINIATURE BUTTONS, SPEAKER VIBRATES RADIO

Sat800 NO BAND SCANNING	**R75** PLASTIC COVERED SMALL SPEAKER

I wish to thank my good friend Pete Gianakopoulos.

81

3.5 Phil's Portable Guide 2005
©2005

1. INTRODUCTION

Passport 2005 reviews 37 portables in 60 pages; we do it in four. Charts are organized by radio size in cubic inches. Negative attributes appear in **gray** and positives are **inverted**. The "Rating" column indicates performance. The "Memories" column reflects only shortwave presets. The "Weight" column is in grams and includes batteries.

2. IS THE RADIO PORTABLE?

We will exclude large radios over 72 cubic inches in volume. The radios below are roughly two to twenty-two times as large as this cut-off value. Two are very expensive and two are poor performers. Notice however that the Satellit-800 is by far the best performing portable.

Radio	Rating	Cost	Size (ci)	Memories	Problems
SW-77	VERY GOOD	$580	142	162	FLIMSY ANTENNA
ATS-818ACS	POOR	$225	198	18	POOR SSB
S-350	POOR	$100	239	0	ANALOG, NO SSB
Satellit-800	EXCELLENT	$500	1641	70	WEIGHS 15 LBS

3. IS THE RADIO DOUBLE-CONVERSION?

We will exclude the single-conversion radios below. Double-conversion is necessary to reduce images. The small radios below are poor performers and have no SSB reception. Units combining digital frequency counter displays with ANALOG tuning experience annoying drift and have no memories. Notice that radios with considerable memories are more expensive.

Radio	Rating	Cost	Size (ci)	Memories	Problems
DE-205	POOR	$15	10	0	ANALOG, NO SSB
KA-105	POOR	$55	11	10	NO SSB
KK-C300	POOR	$40	12	10	NO SSB
KA-818	POOR	$35	12	0	ANALOG, NO SSB
KK-989	POOR	$25	14	0	ANALOG, NO SSB
G1000A	POOR	$50	15	0	ANALOG, NO SSB
E100 (PL-200)	POOR	$100	16	200	NO SSB
CX-CB91	POOR	$15	17	0	ANALOG, NO SSB
KK-S320	POOR	$55	19	5	NO SSB
YB-50	POOR	$80	20	0	ANALOG, NO SSB
M300PE	POOR	$40	22	0	ANALOG, NO SSB
G2000 Porsche	POOR	$80	24	10	NO SSB
KK-E200	POOR	$65	28	12	NO SSB
YB-550PE	POOR	$100	28	200	NO SSB
ATS-404	POOR	$80	39	18	NO SSB
JX-M14	POOR	$30	42	0	ANALOG, NO SSB
E10 (PL-550)	FAIR	$130	42	500	NO SSB

4. CAN THE RADIO HANDLE ADJACENT INTERFERENCE?

Selective fading is common; as is one-sided interference. A radio needs either sideband-selectable SAM or two (narrow and wide) filter bandwidths. In the later case the narrow filter is selected and the radio is detuned up or down by one or two kHz to reduce adjacent interference.

Sony offers four radios with sideband-selectable SAM; however, high cost disqualifies the ones below. Each has few memories, a poor speaker, and no tuning knob. Notice however that the SW-100S is the best performing mini-radio (only 13 cubic inches).

Radio	Rating	Cost	Size (ci)	Memories	Problems
SW-100S	GOOD	$360	13	30	POOR SPEAKER, NO KNOB, CLAM
SW-07	VERY GOOD	$355	25	10	POOR SPEAKER, NO KNOB, CLAM
SW-1000T	GOOD	$450	45	32	POOR SPEAKER, NO KNOB

The radios below were disqualified for having no sideband-selectable SAM and only one bandwidth. SSB reception is either missing or poor. The SW-35 and SW-40 do not come with an AC adapter (~$20).

Radio	Rating	Cost	Size (ci)	Memories	Problems
eTraveller VII	POOR	$100	10	10	NO KEYPAD, NO SSB
ATS-606	FAIR	$120	25	54	NO SSB
SW-35	FAIR	$79	42	50	NO KEYPAD, NO SSB
SW-40	POOR	$120	46	20	NO KEYPAD, NO SSB
YB-80	GOOD	$160	52	18	POOR SSB
ATS-505	FAIR	$120	72	18	POOR SSB

5. ANY OTHER PROBLEMS?

The KA-1101 only has 20 memories while the SW-55 is costly at $360. Both have two bandwidths but no sideband-selectable SAM.

Radio	Rating	Cost	Size (ci)	Memories	Problems
KA-1101	GOOD	$66	23	20	NO KNOB
SW-55	GOOD	$360	59	125	EATS BATTERIES

6. THE TOP FIVE

All five radios below are *recommended*. There are three familiar faces and two small new-comers from Degen. The YB-400PE and DE-1103 have the better "pause and continue" type band scanning. The chart on the next page lists some pluses and minuses for each.

Radio	Rating	Cost	Size (ci)	Memories	Weight (gr)
DE-1102	GOOD	$64	28	133	354
DE-1103	GOOD	$64	32	268	392
SW-7600GR	GOOD	$146	45	100	716
YB-400PE	GOOD	$130	60	40	733
ATS-909	GOOD	$240	69	307	999

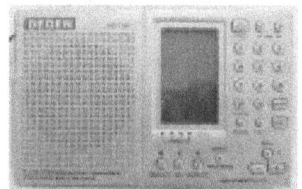 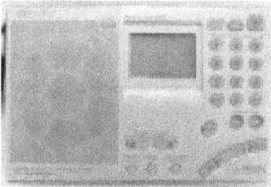 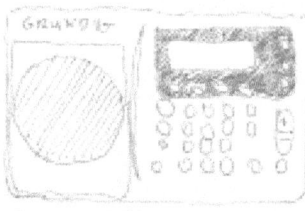 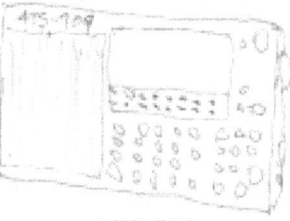

DE-1102   SW-7600GR   YB-400PE   ATS-909

Radio	Pluses
DE-1102	AUDIO, FM
DE-1103	KNOB, GOOD DYNAMIC RANGE, FM
SW-7600GR	SIDEBAND-SELECTABLE SAM
YB-400PE	AUDIO, FM
ATS-909	KNOB, ALPHA TAGS, 40-Hz STEPS

Radio	Minuses
DE-1102	SSB ONLY IN "PAGE 9"
DE-1103	ODD VOLUME, NO 5-KHz STEPS, HIGH CURRENT
SW-7600GR	NO AC ADAPTER, NO EARPHONES
YB-400PE	POOR SSB, NEEDS 6 "AA" BATTERIES [NOT 4]
ATS-909	NEEDS EXTERNAL ANTENNA, EATS BATTERIES

7. THE OVERALL WINNER

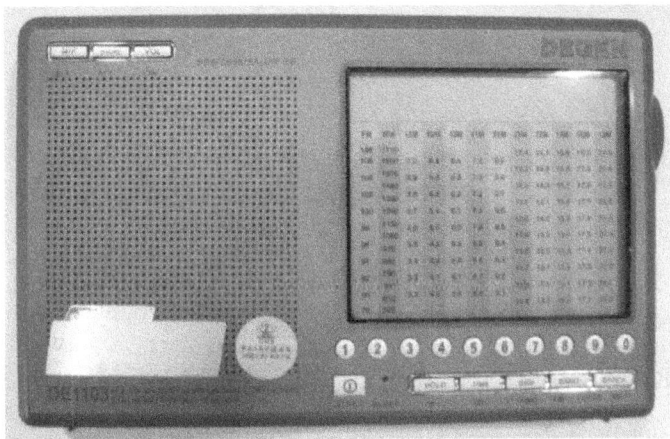

The **Degen DE-1103** costs just $64 on Ebay. Incredibly this includes: the radio, shipping, an AC adapter, a 220V to 110V converter, 4 "AA" NiMH rechargeable batteries, a 30 foot random-wire antenna, earbuds, and a cloth pouch! The radio is only 32 cubic inches, 0.86 pounds, and well built. Features include MW/LW/SW/FM, AM/SSB, two bandwidths (6-kHz and 3-kHz), 268 memories, 17-dB attenuator, clock, dual alarm, sleep timer, keypad lock, simulated analog display, and band scanning. The DE-1103 has a backlight, keypad, tuning knob, wrist loop, and stand. There are earphone, line out, antenna, and power jacks. Coverage includes 100 kHz to 29,999 kHz [AM/SSB] as well as 76 MHz to 108 MHz [FM]. MW and LW reception is via a built-in 5.7" ferrite rod.

To change the volume on the DE-1103 the "VOL" button is touched and then the tuning dial is turned. This is blown into being a major problem but some reviewers. The 5-kHz tuning steps are barely missed because of the handy band scanning feature. Memories can be stepped through via the tuning knob. The buttons have good tactile feel and are backlit for easy usage in the dark. The DE-1103 has greater whip sensitivity and external antenna overload immunity than the SW-7600GR and YB-400PE. The DE-1103 accomplishes this overload immunity by bypassing the 10-dB RF amplifier when connected to an external antenna. The ATS-909 has poor whip sensitivity but does well with an external antenna.

Another great portable is the SW-7600GR with its sideband-selectable SAM-mode. Comparatively, the SW-7600GR offers no AC adapter, no rechargeable batteries, no earphones, no tuning knob, and less than half the memories: in a 40% larger package at 2.5 times the cost of the DE-1103. The YB-400PE and ATS-909 are over twice as costly as the DE-1103, do not include batteries, and are nearly double in size. The YB-400PE has distorted SSB reception; while the ATS-909 battery cover and stand can break easily.

The DE-1103 is small and inexpensive enough to take anywhere. Double-conversion reduces images, digital synthesis reduces drift, a second bandwidth reduces adjacent interference, and good dynamic range allows usage of an external antenna. The tuning knob and simulated analog display are nice features. The radio is capable of listening to hams using SSB. Degen has become a significant player in the portable market.

8. PORTABLE RADIO RECOMMENDATIONS

Distinction	Radio	Cost	Size (ci)	Memories
BEST OVERALL	DE-1103	$64	32	268
RUNNER UP	SW-7600GR	$146	45	100
BEST PERFORMER	Satellit-800	$500	1641	70
BEST MINI-RADIO	SW-100S	$360	13	30

I wish to thank my good friend Ken.

3.6 Tabletop Receiver Notes
ⓒ2003

1. INTRODUCTION

The tabletop receiver landscape is wrought with complicated technical specifications. There are two popular receiver publications and each speaks to different audiences. *Passport to World Band Radio* (Passport) is aimed at beginners and major broadcast listeners. *World Radio Television Handbook* (WRTH) caters to the ham, utility, and DX (distance reception) crowd. Passport values ease of use while WRTH values low-cost performance. All radios, even $4000 ones, have flaws. Electronic equipment design involves compromises.

This is a compilation of data on the 2003 non-professional receiver offerings by: JRC (NRD545), AOR (7030+), Drake (R8B), TenTec (RX-350), ICOM (R75), Yaesu (FRG100B), Palstar (R30), and Grundig (Satellit800). *Tabletop Receiver Notes* will pull no punches while presenting information about receiver features, specifications, and flaws. Discover what rig-owners dislike about their own radios.

2. SHORTWAVE

Shortwave listeners hear public broadcasts. To aid reception by low-performance radios (ex. portables), these signals are transmitted as double-sideband with a carrier (DSBc) at high power. Audio fidelity and ease of usage are important. Some listeners are after utilities (military, aviation, maritime, beacon), hams (amateur radio operators), pirates (illegal transmissions), "mystery" radio (clandestine), DX (distant reception), and QRP (low-power reception). These signals are often transmitted as single sideband (SBB) at low power. Radio features aid DX reception as each "tool" might help "dig" an elusive signal out of the "mud" (noise). It takes a more complex radio to receive SSB transmissions. Most portables offer poor SSB reception. Some people chase longwave (LW) or distant broadcast band (BCB, or regular AM radio) signals. A flexible receiver accommodates changing radio interests. Both AM and SSB are amplitude-modulated emissions. The "AM" mode on a receiver is really "double sideband with carrier" mode.

3. CATEGORIZING

Radios fall into categories. Four receivers are of higher in cost: the NRD545, 7030+, R8B, and RX-350; and four are lower in cost: the R75, FRG100B, R30, and Sat800. The price of the 7030+ is attractive in Europe. The most popular four units are the R8B, R75, R30, and Sat800. Four units are larger in size: the NRD545, R8B, RX-350, and Sat800; and four are smaller in size: the 7030+, R75, FRG100B, R30. The Sat800 is twice the volume of the other large units and 14.6 lbs. The R30 is tiny: half the volume of the other small units and 2.2 lbs. Four radios are more "communications receivers": the NRD545, RX-350, R75, and FRG100B; whereas, four are more "broadcast listeners": the 7030+, R8B, R30, and Sat800. Some of the former four were praised by WRTH and some of the later four were praised by Passport.

4. SPECIFICATIONS

Specifications invite comparison; but consider the source. There can be sample variation, testing inconsistencies, equipment differences, and error. Manufacturers tend to be conservative because they can be held legally to their specifications. Some tests are "apples to oranges". For example, manufacturers often avoid "hard sensitivity" measurements whereby **a 50-Ohm through load is inserted between the RF generator and the receiver** (versus a direct connection). When specifications eclipse those of the manufacturer it could be due to testing a "hotter" sample. If specifications are missing from the manufacturer's literature, it could be because the measured values were poor.

5. FEATURES

Features are "all or none" as a radio either has or does not have a feature. Some items can be crucial to reception and to the enjoyment of the hobby. The RX-350 is the only radio with a bandscope. Two radios offer extended frequency coverage: the R75 (30-60 MHz VHF) and Sat800 (118-137 MHz aeronautical). Three receivers have 1000 or more memories: the NRD545, R8B, and RX-350. The 7030+ has 400, the R75 has 100, the R30 has 100, the Sat800 has 70, and the FRG100B has 52 memories. Both the R30 and the Sat800 can be run off batteries. The R75 has dual pre-amplifiers and a dual-PBT (dual-passband tuning).

The radio with the least number of selectable filters (two) is the R30. Notch filters are used to eliminate heterodynes (single-pitched noises). The NRD545, RX-350, and R75 have automatic notch filters while the R8B has a manual notch. The 7030+ manual notch and noise blanker (NB) are offered as a separate module. Noise blankers stop impulse noise and are missing only on the R30 and Sat800. The FRG100B and R30 have no keypad (for direct frequency entry) and no FM-mode (the FRG100B offers FM as an option). The RX-350 keypad is optional while the 7030+ sports a remote keypad. The R30 alone is missing scan, a squelch, and a clock.

The FRG100B, R30, and Sat800 do not have basic distance reception (DX) tools: no PBT, no RF-gain, and no notch filter. They also lack computer control (RS232). Half the receivers are missing tone controls: the RX-350, R75, FRG100B, and R30. This does not present a significant problem because all eight receivers (except the Sat800 and 7030+) come with an inadequate stock speaker. External amplified computer speakers with tone controls can be bought for ~$25.

6. DSP INTERMEDIATE FREQUENCY

The latest technology is DSP IF (Digital Signal Processor Intermediate Frequency). It allows a plethora of digital filter bandwidths. Chip set costs for the best DSP units are thousands of dollars. The currently used DSP units have inadequate internal ADCs (Analog to Digital Converters) and insufficient processing power. Ultimate rejection and other specifications are poor. DSP technology presents a huge learning curve that each company must climb as they refine their DSP algorithms. The receivers with DSP IF include: the NRD545 and the RX-350.

The R75 is the only receiver of the eight with dual-PBT. This feature allows regular (non-software) filters to select bandwidths from ~400 Hz to ~2400 Hz without the drawbacks of the DSP IF receivers. Placing these filters over one sideband of an AM/SAM transmission allows the PBT to simulate AM/SAM filters of ~800 Hz to ~4800 Hz. The R75 also comes standard with a 6000 Hz filter. Contrary to popular belief, the R75 does not need any of the $70 to $170 optional ICOM filters unless CW (Continuous Wave, Morse code) reception is needed. The R75 uses an AF (audio frequency) DSP to provide both noise reduction and automatic notching.

The R8B is also unique in that it comes standard with an excellent assortment of five LC (inductive-capacitive) filters. This unit has one filter that is being manipulated, by de-Qing the 50kHz filter network, just like the old "4" line. No additional filters are necessary. This combined with its good PBT and phasing allows "simulation" of many bandwidths.

7. SYNCHRONOUS SELECTABLE SIDEBAND

Synchronous detection helps reduce a type of distortion caused by selective fading of the carrier and its sideband components. The SAM-mode (Synchronous AM-mode) maintains synchronization with the incoming carrier. SAM-mode is a good feature and touted by manufacturers to sell radios. Selective fading causes less problem when AM stations are tuned using SSB (so called ECSS or Exalted Carrier Selectable Sideband reception). Synchronous detectors reduce selective fading distortion by reinsertion of the carrier. In a similar fashion, SSB (ECSS) reduces selective fading distortion by using a BFO (Beat Frequency Oscillator) to replace the carrier. SSB (ECSS) is more effective than SAM-mode at reducing strong selective fading because this mode of reception does not have to lock, and maintain lock, onto a carrier.

Some radios use selectable sideband synchronous detection: they maintain synchronization while using a method called "phasing" to select either the upper or lower sideband of a DSBc (double sideband with carrier) AM transmission. The listener simply selects the sideband with the least interference. Using SSB (ECSS) as described above to reduce selective fading distortion necessitates selection of either the upper or lower sideband. Narrow filters are utilized.

Some radios have synchronous detection but do not have "selectable sideband". There is however a trick used to select a sideband under DSB (non-selectable) SAM-mode. Simply engage a narrow (~2.5 kHz) filter in the SAM-mode while applying PBT (or detuning by plus or minus half the filter's bandwidth, ~1.25 kHz) to select a sideband.

Two radios do not offer a SAM-mode: the FRG100B and the R30. The stock R75's synchronous detector does not function properly but can be repaired for $45 or less. The RX-350's SAM-mode functions poorly and firmware updates have been unable to correct the problem.

8. TUNING STEPS AND STABILITY

Some may feel that 1 Hz tuning resolution on SSB is unnecessary because these steps are finer than our hearing can distinguish. The acronym JND stands for Just Noticeable Difference. The human ear has a JND of 1.5 Hz at 500 Hz. What this means is that when presented with a 500 Hz tone, we can tell that a second tone is different when it gets to ~501.5 Hz (1.5 Hz of separation). JND varies with frequency: JND is 2.9 Hz at 1000 Hz, 5.8 Hz at 2000 Hz, and 8.7 Hz at 3000 Hz.

How then can 1 Hz tuning steps be utilized? A trick can be used to center an AM (DSBc) station's carrier down to 1-3 Hz. This, combined with low tuner drift, yields high-performance ECSS (tuning AM stations as SSB) reception. The trick involves using a wide (~6 kHz) filter on SSB and tuning in 1 Hz steps until no "flutter" is heard. After which the normal SSB filter (~2.5 kHz) can be switched back in for listening. This method is fast, accurate, and easy.

Manufacturers omit fine-tuning steps for economic reasons. Using a conventional PLL, fine tuning steps can be achieved in two different ways: 1) multi loop design or 2) a Fractional N Synthesizer. The use of a DDS is probably the easiest way to achieve fine tuning resolution. The DDS (Direct Digital Synthesis) tuning system responsible for LO (Local Oscillator) creation is the most complicated part of superheterodyne receiver design. Only three receivers above have 1 Hz tuning resolution: the NRD545, RX-350, and R75. The other tuning steps are as follows: 7030+ (2.7 Hz), R8B (10 Hz), FRG100B (10 Hz), R30 (20 Hz), and Sat800 (50 Hz). It is not surprising that the most stable tuning systems are also found on the RX-350, R75, 7030+ (TCXO for the synthesizer's reference oscillator), and NRD545.

9. DSP NOISE REDUCTION

DSP (Digital Signal Processor) noise reduction (NR) may significantly aid reception and/or make listening more enjoyable. This feature can protect your hearing while tuning through noisy bands. Some manufacturers omit this feature for economic reasons. DSP technology is costly as it involves hiring DSP hardware engineers and software programmers. Only three receivers offer DSP NR technology: the NRD545, RX-350, and R75. These three receivers are the only triple-conversion radios above. Any radio's audio can utilize aftermarket NR DSP, but the cheapest units are ~$230.

10. QUALITY CONTROL

A quality radio is an investment. Three receivers have quality control (QC) issues. Poor QC presents a unique dilemma when reading reviews: some received "good" units and some receive "poor" units. A professional reviewer should buy "off the shelf" and resist units sent by manufacturers. The R30 has undergone several "silent" revisions in an attempt to "push the envelope". AOR (7030+) and Palstar (R30) both have QC problem histories but also have good customer service channels. Grundig's production of the Sat800 in China has been plagued with poor quality control. Tested Sat800 units can be purchased through R.L. Drake.

11. MODIFICATIONS

Although overlooked by professional reviewers the R75 has an active user community that identified receiver flaws and responded with modifications. These modifications are documented at the R75 Yahoo Group. It is essential to fix the AM AGC, AM synchronous detector, and overall receiver fidelity. Poor AM/SAM performance, the largest complaint levied against the R75, can be fixed for $45 at Kiwa Electronics. Once modified, AM/SAM performance is dramatically improved.

12. RECEIVER SENSITIVITY

Most tabletops have good sensitivity. On paper the R75 and 7030+ are the most sensitive, followed by the R8B and FRG100B, then the NRD545 and RX-350, and lastly by the R30 and Sat800. Actual testing shows the R75 and FRG100B to be similar as well as the R30 and 7030+. Strong signal handling and selectivity are often the more critical aspect of receiver performance.

Usage of an improper antenna can often exploit a weakness of a particular receiver. This can artificially deflate its relative performance during side-by-side comparison. It should be noted that all the tabletops above are cohorts having somewhat "similar" levels of performance. None of these sets is the end-all be-all to radio reception. Professional communications receivers like the Racal 6790 boast impressive specs such as third order intercept points in excess of +30 dBm.

Reception is not magic. A radio is designed to operate within certain designed specifications. View a radio as a "black box" that must be fed proper signals to work. Before blaming the electrical engineers, take a close look at the single most important aspect of reception: the antenna. The wrong antenna can ruin performance. Other aspects of reception include: geographic location, time-of-day versus frequency, gray-line reception, sunspot activity, solar cycles, etc. Propagation varies from minute to minute.

13. THE BIG THREE

Three manufacturers are heavily entrenched in the amateur market: ICOM (R75), Yaesu (FRG100B), and Kenwood. Larger manufacturers do more design research, have lower fabrication costs, and have proven quality control and production lines. They spend significant R&D in high-performance SSB and DSP-related radio technologies. The rumor is that ICOM is licensing Rohde & Schwarz circuitry to build mid-priced ($6000) Mil-Spec receiver gear. These designs will influence their amateur line. The "state of the art" ICOM 756 Pro II uses a 44 MHz DSP.

Grundig is a large manufacturer of consumer portables. Although their Sat800 was elevated to a new category ("Portatop") by Passport, the radio is a very good portable, not a tabletop. Also note that Palstar (R30) is a small manufacturer competing toe-to-toe with the others.

14. GOOD PORTABLE RADIOS

Portables are an inexpensive way to try out the radio hobby. Major AM broadcasts can be received fairly easily. SSB performance is poorer on portables than on tabletops. There are four portables that stand out in terms of price and performance. All four receive shortwave as well as the FM band (88-108 MHz, stereo). The $160 Sony ICF-SW7600GR sports a synchronous detector, runs on 4 "AA" batteries, and has 100 memories. The $240 Sangean ATS-909 (Radio Shack DX-398) sports a real tuning knob, 40 Hz tuning steps, runs on 4 "AA" batteries, and has 306 memories. The $150 Grundig Yacht Boy 400PE sports good audio, runs on 6 "AA" batteries, and has 40 memories. If compactness is critical the AM-only $120 Sangean ATS-606AP runs on 3 "AA" batteries and has 54 memories. Please note that some QC complaints have been levied against the Grundig. Kiwa Electronics offers filter and audio enhancements for the ATS-909.

15. PC CONTROLLED RADIO

The TenTec RX-320 is a good RS232-only controlled radio. Front-end selectivity is fair.

Tabletop Receiver Data

RADIO FEATURE	JRC NRD545	AOR 7030	Drake R8B	TenTec RX-350	ICOM R75	Yaesu FRG100B	Palstar R30	Grundig Sat800
Price in US	$1,800	$1,470	$1,350	$1,200	$530	$600	$500	$500
Price in UK	£1400	£830	£1110	£990	£570	£390	£400	£550
Synch Detector	yes	yes	yes	BROKEN	FIXABLE	-	-	yes
Tone Control	yes	yes	yes	-	-	-	-	yes
Presets	1000	400	1000	1024	100	52	100	70
Keypad	yes	remote	yes	$140	yes	-	-	yes
S-Meter	analog	digital	analog	digital	digital	analog	analog	analog
Tuning Step	1 Hz	2.7 Hz	10 Hz	1 Hz	1 Hz	10 Hz	20 Hz	50 Hz
Tuning Stability	2 ppm	1 ppm	5 ppm	1 ppm	1 ppm	10 ppm	5 ppm	10 ppm
Conversion	triple	double	double	triple	triple	double	double	double
Filters	990	4	5	34	4	3	2	3
PBT	yes	yes	yes	yes	yes	-	-	-
FM-mode	yes	yes	yes	yes	yes	$45	-	yes
Attenuator	yes	yes	yes	yes	yes	yes	yes	yes
Bandpass Filters	yes	-	yes	yes	yes	yes	yes	yes
Adjust Pre-Amp	-	yes	yes	-	yes	-	-	yes
RF-gain Control	yes	yes	yes	yes	yes	-	-	-
AGC	yes	yes	yes	yes	yes	yes	yes	yes
Noise Blanker	yes	$330	yes	yes	yes	yes	-	-
Notch Filter	yes	$330	yes	yes	yes	-	-	-
DSP Noise Red	yes	-	-	yes	yes	-	-	-
Squelch	yes	yes	yes	yes	yes	yes	-	yes
SCAN	yes	yes	yes	yes	yes	yes	-	yes
Clock	yes	yes	yes	yes	yes	yes	-	yes
RS232 Control	yes	yes	yes	yes	yes	$90	-	-
Extra Coverage	-	-	-	-	yes	-	-	yes
Dual-PBT	-	-	-	-	yes	-	-	-
Bandscope	-	-	-	yes	-	-	-	-
SSB Sens (uV)	0.32	0.19	0.25	0.35	0.16	0.25	0.5	0.5
SSB Sel (kHz)	2.4	2.29	2.3	2.25	2.1	2.4	2.4	2.3
Weight (lbs)	16.5	4.9	13	12	6.6	6.6	2.2	14.6
Volume (CID)	944	348	901	788	320	301	163	1641
Inexpensive					yes	yes	yes	yes
Small Size		yes			yes	yes	yes	
Quality	yes		yes		yes	yes		
SAM Mode	yes	yes	yes	BROKEN	FIXABLE			yes
SSB Features	yes	yes	yes	yes	yes			
Ergonomics	yes		yes	yes	yes			yes
Stock Filters	yes		yes	yes	yes			
DX Tools	yes	yes	yes	yes	yes			
DSP NR	yes			yes	yes			
Feature Rich	yes	yes	yes	yes	yes			

RX-350 by TenTec
$1200 ($1340 with keypad)

The RX-350 is known for its digital filters and bandscope display. The radio has had many firmware bugs. TenTec uses flash upgrades to fix the bugs. However, the DSP may lack sufficient processing power to perform all the features, simultaneously. The SAM-mode has difficulty gaining and staying locked, the noise blanker and notch are mediocre, the RF-gain can over-ride the AGC, and the noise reduction alters the audio. The radio can lock up. And a reset erases the 1024 memories.

RX-350 ergonomics pitfalls include: small buttons, a dual knob (for audio and radio frequency amplifiers), quirky tuning step selection, and a $140 optional keypad. The bandscope is small and the display is grainy and ghosts.

The DSP IF provides 34 digital filter bandwidths; but DSP came at a price. The RX-350 is prone to local MW overloading and images. There is hiss and display RFI. The RX-350's specifications (sensitivity, ultimate rejection, IF rejection, dynamic range, and IP3) are not stellar.

TenTec offers a 30-day trial. In many ways, the RX-320 performs similarly: at one-fourth the cost.

Bottom line: The RX-350 uses cutting edge technology. The bandscope and 34 digital filters are tempting; but performance is mediocre. TenTec is a good company that, hopefully, will design another DSP radio with a knob.

NRD545 by JRC

$1800

The NRD545 is known for its quality construction, RF shielded internals, and good power supply (quiet and cool). However, this model cost JRC their "5 star" status in Passport. The unit offers almost 1000 digital filter bandwidths; but the DSP design came at a price. The unit has poor ultimate rejection. The noise reduction and notch are poor. The AM mode sounds bad. And the SAM-mode has difficulty gaining and staying locked,

The NRD545 suffers from DSP-IF related audio problems. The audio has hiss, the tone-control and filters sound digital, and the DSP makes annoying sounds when it overloads. The volume control is mediocre. The display has only 10 Hz resolution while the radio has 1 Hz tuning steps.

Sherwood Engineering has $115 "DSP Protection Filters" that protect against adjacent signal overload.

Bottom line: The NRD545 is a well-built radio with a mediocre DSP IF design. JRC is an great company. Unfortunately, building a DSP IF receiver is difficult. Hopefully there will be an NRD555.

7030+ by AOR
$1470 ($1800 with NB and notch)

The 7030+ is known for its audio quality, strong front-end mixer, and superb specs. The 7030+ can align and center up to six filters. The radio has an IF output and a wireless remote. There is no noise reduction; however, a NB and notch are available for $330.

Ergonomics are poor: minimal buttons and a tree menu system. The keypad is on a wireless remote. The display is small and has RFI. Filter changes can damage the hex screws.

The 7030 line has had multiple quality control issues. An encoder failure problem was corrected in the "7030+3" version. There is encoder wobble. And failures of: buttons, speakers, phone jacks, and power supplies. And power supply noise. Other issues include: slow remote response, SAM filter alignment errors, key bounce problems, AGC table corruption, audio and radio amplifier instability, and J309 JFET and SD5400 mixer failures.

AOR stopped using die-cast metal box shielding around the DDS synthesizer. This reduced PCB stress that caused carrier oscillator failures.

The 7030+ has excellent specs. However, the radio is missing the automatically (PIN-diode) selected bandpass filters that are present in most receivers. It instead relies on a strong front-end mixer. The SSB mode has hiss and only 2.7 Hz tuning steps. The SAM-mode suffers from low-level heterodynes produced by "the harmonic mixing of the DDS signal with the sync VCO". This occurs at the "sync car mixer". For DX, the 7030+ is not as hot as the ICOM and Yaesu units.

Bottom line: The 7030+ is a great broadcast receiver. But the radio has its quirks.

Satellit 800 by Grundig
$500 ($400 reconditioned by R.L. Drake)

The Satellit800 (Sat800) is known for its good audio and Drake-designed sideband selectable synchronous detector. The unit tunes FM (88 to 108 MHz) and the VHF Aeronautical Band (118 to 137 MHz). The Sat800 is popular because of its low cost, good performance, and easy-of-use. This large, 14.6-pound portable is not constructed as well as the other tabletop radios. Battery operation is via six "D" cells.

The unit is missing important DX features: it has no RF-gain, no PBT, and no notch. Other absent features include: no NB, no noise reduction, and no RS232. There is no band scanning (only preset scan of its 70 memories). There is no repair manual.

Ergonomic problems include: a light knob, calculator-like buttons, and a difficult to move antenna.

The Sat800 has a good SAM-mode. The synthesizer has decent phase noise specs but is not as stable (10 ppm) or fine-tuning (50 Hz) as other radios. The unit has some display RFI and tuning knob hash. The Sat800 has the lowest sensitivity of the eight radios.

Sat800 production has been plagued with quality control issues. This includes tuning knob wobble and failures; power supply noise and failures; and dead radio failures. Buy the radio from a dealer with a return policy.

Bottom line: The Satellit 800 is an easy-to-use tabletop alternative. Purchase via R.L. Drake for $400 (reconditioned, tested, and with a one-year warranty).

R30 by Palstar
$495 ($650 with Collins Filters)

The R30 is known for its quality audio, solid front-end, good LW/MW reception, small physical size, and no learning curve. It lacks basic DX tools: it has no RF-gain, no PBT, and no notch. The R30 could use a notch as its wide AM filter allows good fidelity but, occasionally, adjacent carrier leak. The R30 is missing noise-fighting features: it has no NB (noise blanker) and no noise reduction (NR). Other missing features: no RS232, no squelch, no scan, and no clock.

Ergonomics problems include: no keypad, small buttons, a light knob, and vibration from the speaker. At 2.2 lbs the R30 is ideal for portable usage. Five screws (4 external and 1 battery) must be removed to replace the 10 "AA" batteries. There is potential to damage the radio or speaker wires. The unit eats batteries. The R30 has 100 non-volatile memories.

The R30 has good audio but is unarmed to fight noise. There is no SAM-mode to counter selective fading distortion on AM and its 20-Hz tuning steps are not ideal for ECSS reception. The two filter slots cover AM and SSB; but not CW. The display emits some RFI.

External build is good but there have been quality control issues: tuning knob wobble, encoder noise, power switch failures, random resets, etc.

Bottom line: The R30 is the only truly portable tabletop receiver. And it is ideal for LW or MW work. The R30 is a performer; yet lacks many features found on low-cost portables.

FRG100B by Yaesu
$600

The FRG100B, also known as the "FroG", is an underrated radio with a good front-end. The FRG100B has a well lit display, an analog s-meter, and a dedicated knob for quickly tuning through its 52 memories. The FRG-100B's internals are solidly built. The AGC is perfect. Although lacking some features, the FRG100B had **excellent quality control**.

The FRG100B has no SAM-mode, no noise reduction, and no keypad. It is also missing DX features: it has no RF-gain, no PBT, and no notch. Extra modules are needed for RS232 control ($90), FM-mode ($45), and to increase tuning stability from 10 ppm to 2 ppm ($95).

The tuning knob is light, there are only 52 memories, and tuning steps are only 10 Hz. In the FRG100, the AM-mode filters were wide. This can be fixed with Kiwa Electronics' filters. Early units had the poor skirt selectivity (CFW455I filter); however, this was improved in the FRG100B (LF-H2S filter).

Bottom line: The FRG100B is a solid radio. Hopefully Yaesu will produce an FRG200.

R75 by ICOM
$530 ($620 with Kiwa modifications)

The R75 is known for good SSB reception of broadcast (ECSS) and DX signals. This triple-conversion receiver contains many DX tools: RF-gain, dual-PBT, automatic notch, dual pre-amps, slow/fast/off AGC, and FM-mode. And noise fighting tools: NB, attenuator, and DSP NR. And other features: squelch, scan, clock, RS232 control, and 100 memories. There is a learning curve. Reception is to 60 MHz. Cost often includes a free DSP unit: lists for $140.

Ergonomics are good, it has: a durable front plate, dedicated and durable buttons, a keypad, and a sharp and bright display. The rubberized plastic dial has a good feel. The display is not shielded with glass. And pushing the DC adapter in, forcefully, can lead the PCB power connector to break off. The speaker is small and plastic covered. An external amplified speaker is recommended.

SSB and ECSS (tuning AM signals as SSB) performance are good due to the R75's synthesizer having decent phase noise specifications, good stability (1 ppm), and fine-tuning steps (1 Hz). The unit excels at faint signal reception due to many features, and excellent: sensitivity, ultimate rejection, and spurious signal rejection. Dynamic range and image rejection are decent. An optional filter can be placed at two intermediate frequencies. However, optional filters are typically unnecessary (due to the twin-PBT) unless CW reception is the goal.

ICOM botched the AM section of the R75. The AM AGC pumps unless the RF-gain is reduced and the SAM-mode is non-functional. AM broadcasts can still be tuned using SSB (ECSS tuning). Audio fidelity is lacking. Kiwa Electronics can professionally fix the R75's AM AGC, SAM-mode, and fidelity problems for ~$90. Similar modifications (requiring, in total: 8 resistors, 2 transistors, and 2 capacitors) are documented at the R75 Yahoo Group. Post-modification a sideband may be selected via filter under SAM-mode.

Bottom line: The R75 is a feature-rich and well-performing DX receiver whose AM section is flawed. Kiwa can fix this for $90. The stock radio will still hear any signal in SSB mode.

R8B by Drake
$1470

The R8B is known for its good audio, five standard filters, good NB, and the best sideband selectable SAM-mode circuit. Several signal enhancing tools are included: RF-gain, PBT, manual notch, pre-amp, slow/fast/off AGC, FM-mode, analog NB, and attenuator. Other features include: squelch, scan, a clock, and RS232 control. The unit has 1000 volatile memories: they are erased with power loss. The price keeps increasing: the original R8B was $1160.

Ergonomics are good; however, the tuning knob is light, the buttons are calculator-like, and the PBT takes finesse. There have been dial encoder failures. The internal power supply overheats and many run external power. The overheating may account for the ~$200 factory alignments that are can be needed to get used radios running well.

Specs are excellent: sensitivity, ultimate rejection, and spurious signal rejection. Dynamic range and image rejection are good. The synthesizer has good stability (5 ppm) and adequate (10-Hz) tuning steps. There is some display RFI. Five "LC" filters are included; but there is no optional filter slot. The stock CW filter is not ideal for serious CW work.

The R8B's audio is warm. The unit is missing DSP noise reduction and the notch filter is not as good as a DSP notch. Hiss increases when the SAM-mode is engaged. While not a problem for broadcast listening, it can be for DX work in that mode.

Bottom line: The R8B is an easy-to-use and feature-rich receiver. Hopefully Drake will release an R8C or R9.

DISCUSSION

Each receiver above is a winner. Each has survived the natural selection process of a demanding shortwave market by filling a niche. Four units are very popular. The R8B is a high-performance broadcast listener offering ease of use and good distance reception capabilities. The R75 is a high-performance distance reception receiver that excels at broadcast listening after modification. The R30 is a "portable" tabletop with no learning curve. And the Satellit800 combines ease of use and decent performance in a portable.

MY TABLETOP CHOICE

There are three radios in the R75's price range: the FRG100B, R30, and Sat800. All are missing important DX features: RF-gain, notch filter, and PBT. Two have quality control issues and the FRG100B has been officially discontinued. Two are missing synchronous detection, keypads, preamps, FM-mode, and noise blankers. None have DSP noise reduction.

Four radios cost two to over-three times the R75: the NRD545, 7030+, R8B, and RX-350. Two of these receivers use DSP IF: the NRD545 and the RX-350. Looking beyond the multiple digital filters there are some serious performance deficiencies. Check the sensitivity, ultimate rejection, and IP3 values. The RX-350 has image problems and software problems.

The 7030+ has great specs but debatable ergonomics and quality control issues. The specs may have changed since AOR dropped the synthesizer RF-shielding boxes. The unit deserves Passport's "5 star" rating as the publication caters to audiophiles. Price in Europe is good but in North America equipping one similar to the R75 would cost $2030: the 7030+ is $1470, the optional noise blanker and notch filter are $330, and after-market DSP noise reduction is $230.

The Drake R8B is an excellent radio and also deserving of Passport's "5 star" rating. The R8B offers simplicity of operation: five filter-bandwidths and "sideband selectable synchronous detection". The LC filters (Passport rated "good") have shape factors suitable for fidelity. Passport's measured IP3 values (rated "excellent") are decent. Synthesizer stability, resolution, and phase noise values (rated "good") are better on the ICOM. The R75 phase noise specs are superior to the $3950 TenTec RX-340 (as well as sensitivity, ultimate rejection, dynamic range, and IP3). Drake is "made in America" and has a loyal following. The R8B is an attractive unit but pricey.

The R75 is written-off by many due to the botched AM/SAM section. Passport fairly rated the unit "4 & 3/8 stars" stating that the R75 "is a first-rate receiver for unearthing tough utility and ham signals, as well as world band signals received via manual ECSS tuning". The SSB section (capable of receiving AM signals via ECSS) carried the radio. WRTH 2000 rated the R75 and R8B similar in: sensitivity, dynamic range, image rejection, RF intermodulation, IF performance, and audio quality. The R8B had more built-in filter choices where as the R75 ranked better at mechanical design, construction quality, and ergonomics. The R75 was awarded "Best Value Table-Top Receiver" by WRTH. The R75's price dropped by 49% while the R8B's increased by 22%.

Reviews typically fail to speak about the R75 Yahoo Group's modifications. ICOM made electrically minor errors that caused huge performance hits in the AM section of the R75. The fixes involve few components: 1 capacitor for the AM AGC mod; and 4 resistors, 2 transistors, and 1 capacitor for the SAM-mode mod. Audio fidelity complaints were traced to ICOM purposely limiting the fidelity to 3000 Hz. This can be changed to 5000 Hz using just 4 resistors. Kiwa Electronics will professionally perform these modifications for $80: The $45 Synchronous Detector Upgrade and $35 High Fidelity Audio Filter Upgrade. The R75 sells for as low as $450 with the DSP unit included.

Other modifications are available: an ECSS volume increase, an inexpensive Murata filter (15-element ceramic; bandwidth of 4.5 kHz; shape of 1.44; ultimate rejection of 80 dB); as well as Kiwa Electronics' audio enhancement, MW attenuator bypass, and 3.7 kHz AM filter. The bottom line is that ICOM has realized the technological limits at this price point.

I would like to thank my good friend Pete Gianakopoulous.

3.7 Mid-70's Panasonic Portables

In the mid-1970's Panasonic had a lineup of radios. They were no longer the heavy boat anchors, containing tubes. They were not travel radios with small tuning knobs. And they were also not modern (1980's forward) communications receivers with phase-locked loops. What Panasonic created was a line of big-speakered, large-knobbed, linear-capacitor-tuned portable radios. The smallest, the RF-2200, is roughly 20.7 times the volume of a PL-380. The RF-2600, RF-2800, and RF-2900 are 82% larger than even the RF-2200. And the RF-4800 and RF-4900 are 5.48 times the volume of the RF-2200. The RF-2200 ran on four D batteries, the mid-sized of this lineup ran on six D batteries, and the full-sized radios ran on eight D batteries. The RF-2200 was very popular. While the RF-2800 and RF-4800 are more rare. Unfortunately, both have a red LED display (versus a fluorescent blue one) that works only on the shortwave bands (not on MW or FM).

All of these Panasonic radios contained similar features, including: a large speaker, a signal-strength meter, a large tuning knob, and an analog display. There were controls for volume, bass, treble, and narrow/wide filters. Each had a power switch, a display light, a battery check, a BFO (called an AM-CW/SSB switch on high end models), and an RF gain knob (called a local/DX knob on upscale models). Also included were audio and recorder output jacks. All models, except the RF-2200, had an MPX out jack. Each radio covered AM, FM, and SW. Shortwave coverage was, at least, 3.9 to 28 MHz. With the higher end models having extended 1.6 to 30.0 MHz coverage.

The RF-2200, the entry model, is an important analog-only radio. The RF-2200 appears to have been created to lure consumers into the higher priced models. The RF design of the radio is excellent on all bands; but not without its compromises. In Panasonic's lineup, the RF-2200 was the only radio without a digital "frequency counter" display, without a BFO pitch knob, and, yet, included a large, rotatable, ferrite rod for superb MW reception. On shortwave, the RF-2200 contains two crystal calibration switches; whereas, the other models use a shortwave calibration knob. The RF-2600 has no FM AFC on/off switch. The RF-2900 has a digital display on/off switch and some have a SW preselector (band switch and tuning) on/off switch. The preselector should remain off while band scanning. The preselector stops unwanted energy from getting to the first mixer. The RF-2900's that have a preselector also have a longwave (LW) band and FM with AFC only. Each radio, except the RF-2600, has a slow-or-fast tuning knob. The slow setting tends to be helpful on shortwave. On shortwave, the RF-2200 splits SW into six bands. Whereas the RF-2600 has four bands. The RF-2800 and RF-2900 have three bands. And the RF-4800 and RF-4900 have eight bands. The RF-4800 and RF-4900 have two tuning knobs: one for AM/FM/SW1 (a single speed knob) and one for SW2 through SW8 (a dual speed knob). Other unique features include: dual analog displays, an antenna trimmer, an automatic noise limiter (ANL), and three separate antenna hookups (AM-FM-SW). Panasonic basically created portable, transistorized "boat anchor" type receivers. Other manufacturers had similar offering but most are scarce and costly now.

In many ways, a radio is a radio, is a radio. Any modern radio (super-heterodyne, even single-conversion) can be used for DX work. A single-transistor regenerative radio or crystal set can also be used: it just takes more work and skill. Every $400+ communications receiver that I have ever owned was sold: even my ICOM R75. Remaining are the homebrew sets and other select radios, including several Panasonic radios. A key factor in the enjoyment of radio is tuning feel. Panasonic's mid-70's lineup has a winning combination of: a big speaker, a signal-strength meter, a big tuning knob, and gear-driven, frequency-linear capacitors. The capacitors tune both the input tank and the first mixer's local oscillator. Modern radio design is centered around omitting the physical component (ideally a ceramic-standoff, air-variable, frequency-linear, gear-driven, dual-capacitor) that, in my opinion, is essential to the enjoyment of the radio hobby. An even more, electrically-critical component of both tanks (input and LO) is the inductor. The Q of a non-air capacitor is sufficient; however, the Q of the tank's inductor is often a limiting factor. The Hellenized Sony ICF-S10MK2 (written about elsewhere) is a radio in which a lowly $11 Sony AM/FM portable is converted to shortwave via a LO formed using an Amidon inductor and a high-dollar, signal-generator, air-variable capacitor. Even with a non-tuned input, the results are satisfying due to tuning feel. This type of set quickly makes one realize that the modern radio is designed to help manufacturers, not consumers. And for this reason, old radios will always have a strong allure.

RF-2200 by Panasonic

In ~1978, Panasonic released several radios. The RF-2200 was at the top of their portable radio lineup and at the bottom of their communications radios (ex. RF-4800). Cost of the RF-2200 was $150 (~$650 in 2017 money); while the RF-4800 was $450 (~$1950 in 2017 money). The RF-2200 (using 4 D batteries) and the RF-4800 (using 8 D batteries) have similar designs. Both use the PC1018 AM/FM chip and a 2.5 Watt audio chip (HA1329 in the RF-2200 and LA4201 in the RF-4800). Two major differences is that the RF-4800 includes a more complex oscillator (to feed the first mixer) and a LED frequency counter display, on SW. The RF-4800 also has an automatic noise limiter, separate MW-SW-FM antenna hookups, and more frontend tuning. Overall, the Panasonic RF-2200 gets everything right: it is fantastic on MW, FM, and SW. The left side includes two jacks: recorder and headphone outputs. The top includes FM whip and MW rotating antennas. The back includes a 120V connection, AM and FM antenna hookups, and a battery compartment. The front, left, of the radio includes: an on/off switch, a temporary light, a 455-kHz BFO, and a speaker. Also a signal meter (that moves to the left), FM AFC or AM narrow/wide filter, two calibration switches, base, treble, and volume controls. And a display, tuning knob, and fast/slow tuning switch. And, to the right, a SW band selector, FM-SW-AM selector, and MW/SW RF gain control. This radio is a marvel of both mechanical and electrical engineering. The design includes a frequency-linear variable capacitor, a no-backlash gear drive system, and an accurate frequency readout.

The RF-2200 is hot on MW. It includes a ~8.00" space-wound, rotating, ferrite antenna. This is tuned and feeds a gain-controllable JFET RF amplifier setup in common-drain configuration (with chip AGC action). This provides substantial current gain and input impedance. On MW the radio is a single-conversion design with an IF of 455-kHz. There is +10 dB of RF gain, +61 dB of chip IF gain, +5 dB of mid-filter IF gain, and +49 dB of audio gain: about +125 dB total gain. Intermediate frequency filtering comes from four tuned LC "cans" and a switchable 455-kHz ceramic filter.

The RF-2200 is hot on FM. The FM chain consists of a BPF, tuned JFET amplifier (common source, voltage and current gain), mixer, LC-filter, ceramic filter, IF amplifier, ceramic filter, IF amplifier, and PC1018 chip with a third ceramic filter. FM is single-conversion and the IF is 10.7 MHz. For DX: turn the AFC off and use the slow tuning. This radio can hold its own against radios with Silicon Labs' chips. On FM, a RF-2200 is more fun to tune than, for example, a PL-380.

The RF-2200 is hot on SW. Here it is double-conversion with 1.985-MHz and 455-kHz intermediate frequencies. The tuned frontend and 455-kHz filters are the same as on MW. The PC1018 chip is fed a fixed oscillator. The first mixer is followed by two LC filters. The key to DX on SW is to attach a 50 foot antenna and set the RF gain very low (~2). Warning: feed the antenna through a large (0.022 uF) RF capacitor to avoid direct current from entering the RF-2200. Same for the RF-2600, RF-2800, RF-2900, and GE 7-2990A. Extend the antenna in AM/FM mode. Also, drop the RF gain control when: 1) using the BFO (it must be the larger signal), 2) listening to a strong MW/SW station, and 3) general listening, to reduce each band's noise to "2" on the meter (tune to a location without a station and drop the gain until the meter reads on its lower end).

The "Holy Grail" RF-2200

The question that arises is: "What makes the Panasonic RF-2200 such a winner for MW DX?" The obvious reasons, to the consumer, are: 1) a rotating MW antenna, 2) the strong sound of 2.5 Watts of audio, 3) an accurate, linear, analog, frequency readout, 4) excellent tuning feel from a hefty, frequency-linear tuning capacitor, and 5) a good sounding wide filter and an optional narrow bandwidth. Studying the schematic and internals reveals a less obvious list of positives, including: 1) 40 out of 56 or 71% of the ferrite's turns are space-wound for enhanced directionality, 2) use of a sizable ~8.00" ferrite antenna rod, 3) sending the tank input to the high-impedance gate of a n-JFET RF amplifier, 4) use of common drain configuration for added input impedance and high current gain, 5) a controllable (BJT) variable-current drain for the RF amplifier, 6) a ground-shielded RF input run, 7) four inductive-capacitive cans in the 455-kHz chain, 8) a BJT amplifier after two IF cans to preserve the signal, 9) a hefty variable capacitor ground strap, 10) series inductors and parallel capacitors in the power chain to dump RF to ground, and 11) +125 dB of overall gain. Some of the surprising aspects of the design include: 1) it is single-conversion on MW, 2) it is totally chip based except for the added RF amplifier, and 3) there is a fairly ugly BJT PNP switch that selects between a variable MW LO and a fixed SW LO, each heading to the chip. The bottom line is that the RF-2200's strong MW performance is because of attention to details early in the signal chain. Panasonic's engineers did an excellent job of obtaining a good signal (via a well-wound, long ferrite) and immediately current amplified that signal. The rest was fate, even for a chip radio. And where did thousands of other radios drop the ball... the answer: very early.

RF-2600 by Panasonic

This section compares the RF-2600 and RF-2200. The RF-2200 is smaller, more dense, uses four D batteries, has six SW bands, has dual speed tuning, and employs two crystal SW calibration markers. While the RF-2600 is larger, feels hollow, use six D batteries, has four SW bands, has single speed tuning, and employs a SW calibration dial. The RF-2600 has no FM AFC switch; however, it has an MPX OUT jack, a BFO pitch knob, and a green digital frequency readout. The tuning on the RF-2600 feels slow. Using the "slow" tuning speed, it takes the RF-2200 about 18.9 turns to tune the MW band. Using the "fast" speed, it takes exactly five turns. It takes the RF-2600's single-speed tuning knob about 15.4 turns to complete the task. The RF-2600's dial spins easier and has a small crank that aids in rapid spinning. The RF-2600's tuning speed is a compromise; however, the digital frequency readout is a great feature. Ironically, modern analog radios with a digital frequency counter display are typically junk. However, in the mid-1970's, when this was the state of the art, Panasonic offered a superb radio with a frequency counter. The stock RF-2600 is weaker on MW than a RF-2200; but a radio shack loop evens the playing field.

RF-2900 by Panasonic

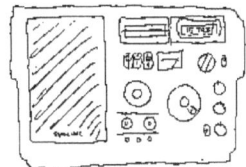

This section compares the RF-2900 and RF-2600. Both radios use the TA7208 audio amplification chip (the RF-4800 and RF-4900 use the LA4201) The RF-2600 has four shortwave bands (aiding its single-speed tuning); whereas, the RF-2900 has three. The RF-2900 has a FM AFC on/off switch, a left side (versus back) power cord, a right side (versus left) extendable antenna, and handles. The big difference is its dual speed tuning knob: pull for slow and push for fast. Fast on the RF-2900 spans the band in ~3.5 turns; while slow takes ~48 turns. The RF-2600 is uniquely styled with its rounded speaker grill and rounded fluorescent display window. The fast tuning on the RF-2900 and RF-2200 have better tuning feel, in my opinion, than the RF-2600; which feels slow. However, the slow tuning on RF-2900 and RF-2200 seem too slow. My homebrew radios span 4.7 MHz to 10 MHz in 2.5 turns. This is 50% faster than even the RF-2900's "fast" setting.

This section compares the internals of the mid-70's Panasonic radios. Except for the RF-2200, each radio has a divider, a LSI counter chip, and a digital display. The RF-2200, alone, uses the AGC output of the PC-1018 chip to change the RF gain setting. Except for the RF-2800, each radio has an audio amplification chip (varies). The RF-2200 utilizes the PC-1018 RF chip: a mixer and an IF amplifier. The RF-2600 and RF-2900 utilize three RF chips: the AN7212 (oscillator and mixer), the AN7210 (mixer and IF amplifier), and the AN7211 (AM/SSB detector and meter circuit). The RF-2200, missing these three chips and with an accurate analog frequency readout, was likely both electrically and mechanically hard to design. The RF-4800 and RF-4900 utilize discrete components (ex. ~32 inductor cans) as well as the PC-1018 input chip. They use triple tuning for shortwave (3 MHz to 30 MHz). There is a tuned circuit both before and after the RF amplifier that feeds the first mixer. However, on MW, FM, and from 1.6 MHz to 3.0 MHz there is a setup similar to all the other radios: a single, tuned circuit before the RF amplifier. They uses single-conversion for MW, FM, and SW1 (1.6 MHz to 3 MHz). Their major design enhancements aid in SW listening.

On FM, the RF-2200, RF-4800, and RF-4900 are similar, using 3 ceramic filters. The RF-2900 cut corners, but has 3 ceramic filters. The RF-2600 cut more, using a common base BJT amplifier instead of a JFET and 2 ceramic filters. The GE-7220A is a similar radio offered by General Electric, but designed by Panasonic engineers. The GE-7220A uses an FM frontend chip and only 2 ceramic filters. All radios are protected on FM. On MW, the RF-2200, RF-4800, and RF-4900 use a 1-chip solution. The other radios use a 3-chip solution: MW is convoluted due to bypassing one chip's mixer. The RF-2200 and RF-2900 are protected by what amounts to a gimmick capacitor. The RF-4800 and RF-4900 are inductively (strongly) protected. MW is fully protected on the RF-2600 and GE-7220A by not being hooked to their external antenna. On SW, the design exceptions are the RF-4800 and RF-4900. The RF-2200, RF-4800, and RF-4900 are 1-chip solutions. They use a dual NPN (2SA838) first mixer. The other radios use chip solutions (oscillator/mixer). In SW mode, avoid touching the antenna of the RF-2200, RF-2600 (SW1 is protected), RF-2900, and GE-7720A. An external wire hooked to the antenna port is safe. DO NOT clip a wire to the built in rod or the JFET could get damaged. The RF-4800 and RF-4900 are totally protected (inductively) on shortwave.

The best of these six, mid-70's Panasonic radios are: the RF-2200 (MW DX, compact, a mechanical marvel), the RF-2600 (digital frequency counter display, rounded stying), and the RF-2900 (digital counter, dual speed tuning knob). I recommend the RF-2200. And either the RF-2600 or RF-2900. With careful shopping, these three can be had for ~$100 each. Some sell nearer to $350. I do not recommend the RF-2800 and RF-4800 due to their lack of frequency readout on MW and FM. The RF-4900 is a large radio: it's a unique, plastic, transistorized, pseudo- boat anchor. Some competition to Panasonic's lineup are Sony's ICF-5500 (single-conversion), ICF-5800 (single-conversion), and ICF-5900W (crystal calibrator). None have a digital frequency counter.